Lean Six Sigma

A Practical Guide for Getting Started with Lean Six Sigma along with How It Can Be Integrated with Agile and Scrum

© **Copyright 2020 - All rights reserved.**

The contents of this book may not be reproduced, duplicated, or transmitted without direct written permission from the author.

Under no circumstances will any legal responsibility or blame be held against the publisher for any reparation, damages, or monetary loss due to the information herein, either directly or indirectly.

Legal Notice:

This book is copyright protected. This is only for personal use. You cannot amend, distribute, sell, use, quote, or paraphrase any part of the content within this book without the consent of the author.

Disclaimer Notice:

Please note the information contained within this document is for educational and entertainment purposes only. Every attempt has been made to provide accurate, up to date, and reliable information. No warranties of any kind are expressed or implied. Readers acknowledge that the author is not engaging in the rendering of legal, financial, medical, or professional advice. The content of this book has been derived from various sources. Please consult a licensed professional before attempting any techniques outlined in this book.

By reading this document, the reader agrees that under no circumstances is the author or publisher responsible for any losses, direct or indirect, which are incurred as a result of the use of the information contained within this document, including, but not limited to errors, omissions, or inaccuracies. Any icons used in this book are the property of Freepik and received through a Basic License. Besides, files from the Wikimedia Commons are used. Commons is a freely licensed media file repository. All credit belongs to the authors/creators.

Contents

INTRODUCTION 1

CHAPTER 1. WHAT IS LEAN PROJECT MANAGEMENT? 5

    THE ORIGINS OF LEAN 8

    LEAN PRINCIPLES AND BENEFITS 10

    LEAN BENEFITS AND WASTE 15

CHAPTER 2. SIX SIGMA: TOOLS, ROLES, AND CONCEPTS 19

    THE ORIGINS OF SIX SIGMA 23

    KEY ELEMENTS AND ROLES 24

    GETTING STARTED: THE STEPS AND TOOLS 27

CHAPTER 3. LEAN + SIX SIGMA = LEAN SIX SIGMA 33

CHAPTER 4. TRAINING AND CERTIFICATIONS: WHAT DO I NEED? 38

CHAPTER 5. THE LEAN SIX SIGMA PROCESS: DMAIC VS DMADV .. 41

    DMAIC 41

    DMADV 45

DMAIC VS DMADV 48

CHAPTER 6. DEFINE: PROCESS MAPPING AND CUSTOMER VOICE 51

What is the Define Phase? 51

Process Mapping: SIPOC 53

Voice of the Customer: Identifying and Understanding Customers ... 58

CHAPTER 7. MEASURE: PROJECT WHY'S, DATA, AND DEFECTS 61

What is the Measure Phase? 62

Data Types and Data Collection 64

Identifying Project Ys 65

Variations and Defects 66

CHAPTER 8. ANALYZE: FINDING POSSIBLE AND ROOT CAUSES 69

What is the Analyze Phase? 70

Value Stream Mapping: Identifying Waste Causes 72

The Five Whys Method 73

Hypothesis Testing 75

CHAPTER 9. IMPROVE: GENERATING SOLUTIONS 78

What is the Improve Phase? 78

Steps Toward Improvement 79

Cost-Benefit Analysis 80

Solution Parameters and Generating Possible Solutions 83

CHAPTER 10. CONTROL: SUSTAINING IMPROVEMENT 86

What is the Control Phase? 87

The Control Plan 88

Control Chart 89

Mistake Proofing: Poka Yoke 91

CHAPTER 11. LEAN SIX SIGMA WITH AGILE AND SCRUM 95

CHAPTER 12. MISTAKES TO AVOID IN LEAN SIX SIGMA 101

CONCLUSION 105

Introduction

For every business, there's always room to do better, always room to be more productive, and to make more profit. And there's always room to waste less time and resources. Although many business owners possess excellent skills in their various industries, many are oblivious regarding properly managing projects and achieving the best results. Thus, to fill this gap, they seek out new human capital to help them do that. Specifically, they are on the search for talented project managers. These are project managers who aren't only aware of the traditional way of performing projects, but who also have experience in the new methodologies too. They're aware that most projects are filled with waste and inefficiencies. How do they know this? They're equipped with the right knowledge about Lean Six Sigma.

So, what is Lean Six Sigma all about? Lean Six Sigma is a globally proven and applied methodology that delivers a demonstrable and sustainable improvement of processes and organizations. The focus is on matters *that matter to customers*. This methodology leads to simultaneously reducing costs, increasing customer satisfaction, and shortening lead time. This involves investing in human capital and making optimal use of people's knowledge and skills for cost-effective and successful projects. Lean Six Sigma offers a framework with which organizations can achieve concrete results with continuous improvement in a structured manner. Implementing Lean Six Sigma gives an organization a goal-oriented approach by translating the strategy into tangible results. Lean Six Sigma projects ensure a sustainable improvement of the operating result. Business problems are solved better and permanently. The return on investment (ROI) of a Lean Six Sigma process varies - but is usually multifold of what was invested.

The basis for Lean Six Sigma is data and facts. This concerns both making the results measurable, providing insight into the real wishes of the customers, reliably measuring current performance, and objectively testing the root causes. In this way, you can tell if the solutions being implemented have the desired effects. Both Lean and Six Sigma started as separate methodologies. Lean strives for more transfer and value generation, while Six Sigma strives for stable and effective processes. In combination, they reinforce each other and are entirely complimentary. Organizations are confronted with rising costs and increasing competition on a day to day basis, but using the Lean and Six Sigma combination, you can combat these problems and grow your business. There are great benefits found in this methodology, such as an increase in profits, improved efficiency and effectiveness, lower costs, and support for employees to develop their skills and knowledge.

It's deplorable that many project managers spend a lot of time and energy working toward ill-defined project goals. These goals don't matter to the organization and are poorly stipulated because no time was taken out to define what's really important to customers and, thus, the organization. To remain concise, many project managers fail to understand the intricate difference between efficiency and effectiveness. When an organization is efficient, this means getting as much done as possible in as little amount of time. But the mantra of "doing more with less" is pointless without direction, i.e., effectiveness. When an organization is effective, this means that they are moving forward on the right path and *know their destination*. While traversing this right path, successful organizations aim to do more with fewer resources, like time and money.

All of this talk about Lean Six Sigma may sound too good to be true for you as a new or aspiring project manager. However, it's real, and many organizations around the world are proving it to be an effective and powerful program. Although much has been written about project management, too much information is too generic and not applicable in today's ever-changing world. Thus, this guide was written to cover all aspects and processes of Lean Six Sigma in a secure and up-to-date manner. Besides, this book contains simple language and no fluff, so every beginner can follow along without losing track of its main points. Due to the lack of practical guides on the topic, my aim in this book is to share with you as many practical examples, case studies, and expert advice you need to start applying the concepts right away. The examples, case studies, and specialist advice are distilled from my experience managing projects and from other very successful project managers. So, even more seasoned project managers can benefit from this book. All this results in a comprehensive book on the topic; you don't need any additional material to get started. Finally, the guide is also suited for those who want to pursue a Lean Six Sigma Certification - or aren't yet sure if they're going to acquire one.

This book consists of two primary parts, namely: The Essentials and The Process. Within these recipes, we find various ingredients needed to make the "correct meal." Without a guide, without the right elements, it's impossible to complete a project successfully. The first part, The Essentials, is covered in Chapters 1-4. In Chapter 1, I'll explain more about what Lean Project Management is all about. By reading this chapter, you'll learn more about its origins, processes, and principles. In Chapter 2, you'll have a solid understanding of the various tools, roles, and concepts within the Six Sigma framework. This is related to Chapter 3, wherein I explain the

Lean Six Sigma methodology and benefits in beginner-friendly language. To finish this first part, Chapter 4 gives you a view about training and certifications and answers the question of whether you need one.

The second part, The Process, is covered in chapters 5-12. This part is all about the practical process of Lean Six Sigma. In chapter 5, we'll delve deep into the Lean Six Sigma Process and what DMAIC and DMADV are all about. Afterward, Chapters 6-10, are all dedicated to the various phases in Lean Six Sigma, namely the Define, Measure, Analyze, Improve, and control phases.

Furthermore, in Chapters 11, I'll give you more information about the relation between Lean Six Sigma, Agile, and Scrum. Finally, in Chapter 12, you'll learn more about mistakes to avoid when implementing Lean Six Sigma. I'll explain the secrets of how to properly combine Lean Six Sigma with the very practical Agile approach to prevent many of these mistakes. This will set you up for more success in projects. So, are you ready to uncover these secrets and grab the necessary techniques? Let's go!

Chapter 1. What is Lean Project Management?

The rise of new technologies has brought much dynamicity in the work we do. Because of the dynamic environment organizations function in today, we not appropriate ways to manage this dynamicity. Traditional project methodologies, like the Waterfall Method, aren't very feasible with rapidly changing customer needs and wishes. Fortunately, with the rise of software development, great minds came together to craft an approach to better manage projects. This resulted in approaches like Agile and Lean. Agile brings about many benefits, consider the following:

- **It results in shorter lead times.** In the past, an IT project consisted of a lengthy, usually unrealistic planning. If one part was canceled or if the planning had to be revised for another reason, this would entail a lot of trouble. Often, the result was very different from what was intended. Then the whole project - or significant parts of the project - would start again, often with a considerable lead time as a result. The Agile approach ensures considerably less administration because you only have to adjust the following sprint (defined time to deliver a product increment) instead of the entire project planning. This makes the lead time of the total project a lot shorter.

- **It paves the way for better customer satisfaction.** With traditional project management methodologies, the customer was involved in the kick-off and delivery. However, the process in between determined whether the result was according to the original plan. In Agile, the customer remains involved throughout the entire process and provides input and feedback during every sprint. Thus, it is continuously checked whether all stakeholders are on the same line. This ensures higher customer satisfaction.

- **It results in better, higher quality projects.** If you work with Agile, you gradually process feedback and customer wishes in your project. In the scrum method, a (partial) product is delivered after every sprint. Everything is also tested, and errors and deviations are quickly discovered. You can guarantee that the end product not only resembles the original plan but also works properly. With the Agile approach, you increase the quality of your solution. You provide the customer with real added value.

- **It lowers risks and costs.** Agile is a type of risk management because there's always room to readjust. The constant feedback received from customers and the higher quality of the delivered incremental products reduce the risk of errors and loss of resources.

- **More teamwork.** Working incrementally towards a common goal results in close cooperation between all disciplines, such as business analysts and software developers. There is complete transparency about what each team member does and at what time. There is more focus, no unnecessary meetings, and you see a ton of quick wins resulting from the team's efforts. This ensures more satisfaction and motivation for the entire team.

The Agile approach has brought various methodologies for better managing projects, such as *Lean*. Lean means slim. By employing Lean, organizations and businesses, make their processes "slimmer." This means that all wastage, "the unnecessary excess fats," are eliminated. With *waste*, we refer to elements for which the customer doesn't want to pay or wait; things that offer no added value to the customer, such as when a manufacturing worker needs to walk to the other side of the factory to grab some equipment. With Lean, everyone within the company can respond quickly and perform optimally. It consists of these key elements:

- Make sure to reduce stock
- Hyper-focus on reducing the lead time
- Strive for as much flow and least "hiccups" in the process as possible, creating flow in the process;
- Create a sense of *Pull*;
- Always aim and go for perfection.

The central idea within the Lean methodology is to put customers first and create more value for them with minimal effort. This approach helps you achieve increased customer satisfaction, employee satisfaction, and continuous quality improvement. Applying the Lean methodology brings your processes in an optimal form to reach peak performance and not get "exhausted." With Lean, a company produces both more efficiently and more sustainably. A more efficient production process that has been optimized through Lean uses fewer materials and energy per production unit. This is due to far less downtime and fewer decoupling stocks.

Although Lean started in the automotive industry, today its used in all kinds of industries, from IT companies to clothing manufacturers. If you decide to implement Lean Management in your business or organization, awareness, and acceptance of your team members is essential to reach the desired outcomes. These elements will be explained in more detail later on. However, remember that no process, organization, or culture is the same. There's always a certain degree of adjustments needed to make the best use of Lean. Usually, it's taking one step back to take two steps forward. Let's now take a step back and learn how Lean started before we delve into further intricacies!

The Origins of Lean

Lean is, by far, the most used methodology to improve processes. But what are its origins? How did it come about? Who used it first? To answer these questions and more, we need to take a step back in history, back to the end of the nineteenth century. An entrepreneur and Japanese inventor, Sakichi Toyoda developed a mechanical loom. That kicked off a textile industry revolution in Japan, and in January 1918, he started his company, the Toyoda Spinning and Weaving Company. His son, Kiichiro Toyoda, helped and, although there were many setbacks, by 1924, his dream came alive, and he could finish constructing an automatic loom. In 1926, Toyoda Automatic Loom Works was opened.

To do all of this, Toyoda made good use of the Jikoda Principles, the main one of which stood for incorporating quality automatically into the main objective, i.e., production, free of defects, and removing any redundancies. Later, Sakichi was to make a change to the name of his company, calling it "Toyota."

The family didn't stop there, though. As an inventor, his son made many visits to the US and Europe in the 1920s, and it was there that he met the automotive industry. When Sakichi Toyoda sold the patent to the automatic loom, his son used the money to set down the foundations for a new company. In 1937, the Toyota Motor Corporation was born, and Kiirichi began to produce the company's first vehicles, using General Motors parts. One of his most important legacies was the Toyota Production System. Kiichiro's "just-in-time" philosophy - production of only the exact amount of parts already ordered striving for minimal waste - was an essential factor in the development of this system. The Toyota Production System was slowly but surely being used by more and more car manufacturers throughout the world.

After the Second World War, a cousin of Toyoda's, Eiji Toyoda, who now also worked for Toyota, visited Michigan. Here he attended the Ford factory in Dearborn to study their method of production. On his return, he discovered that the way Ford manufactured cars wasn't feasible in Japan, because of the smaller market. Thus, the company searched for an alternative way of manufacturing, which turned out to be utterly different from Ford's highly effective mass production system. When Japan began to pick itself up after the industrial chaos of the war, Toyota became the largest Japanese automobile manufacturer, having a current market share of around 42%. At the end of the fifties, Toyota delved into various regions. The first Toyota Crown vehicles arrived in the US in 1957. And in 1965, with models like the Toyota Corolla, the company started to build a reputation and sell numbers that could compete with those of local producers.

This was never possible without the alternative the company found. They shifted their focus on an entirely new way of producing with attention to development, manufacturing, delivery, assembly, and of course, labor. This new approach is named the Toyota Production System. The following four Ps can describe the basis for this system:

- **Problem-solving is first and foremost.** While producing cars, we have to learn and continuously improve the processes to gain the best results.

- **People & Partners are critical to success.** Without the right partners, processes cannot run. If a partner slacks down, the whole process slacks down. Therefore, great partners are needed who give great attention to transparency: hard work, and collaboration.

- **Process(es) run the production.** A system is based on a multitude of processes. Problems in one process can have a direct effect on another process. Thus, dealing with them in a professional or entrepreneurial way is necessary.

- **Philosophy keeps us on track.** Without the right philosophy, nothing worthwhile can be achieved. Thus, focusing on instilling long-term thinking, great collaboration, a clear mission, and vision in every team member is critical.

The term Lean was first championed in the west by John Krafcik in his paper *Triumph of the Lean Production System* (1988). In the paper, Krafcik gives more insights into productivity and quality levels in the auto industry. Before his research, people in the auto industry believed that an assembly plant's location determines productivity and quality levels. He concluded that plants operating with a more *Lean* approach could produce motor vehicles faster while retaining high levels of productivity and quality. In 1990, James Womack, Daniel Jones, and Daniel Roos wrote the book *The Machine That Changed the World*. It discusses a study they conducted into the difference in effectiveness between various car manufacturers. They found that Toyota was first and foremost in the global automobile branch, because of the Lean production principles they applied. These principles were later detailed in a paper written in 1996 by James Womack and Daniel Jones, called *Lean Thinking: Banish Waste and Create Wealth in Your* Corporation. The principles mentioned in the article were detailed further in their book with the same title. By now, I hope I made you more curious about these principles that made Toyota extremely successful in their industry. Let's take a closer look.

Lean Principles and Benefits

Now that we've taken a look at the origins of Lean, what are these principles all about I mentioned previously? According to James Womack and Daniel Jones, there are five principles of Lean Thinking, namely:

- Specify Value;

- Identify the Value Stream;
- Flow;
- Pull;
- Pursue Perfection.

Lean Management focuses on removing waste from processes. This results in working smarter, not necessarily harder, to better serve your customers. It doesn't matter in what industry you'd like to apply Lean; it's beneficial for pretty much every industry and company size. To turn your organization into a Lean organization, understanding and to apply the principles mentioned above helps you build a strong foundation. Let's take a closer look at each principle.

The first principle is defining or specifying *value* by putting first things first, i.e., your customers. A business is nothing without customers, depending on them and growing because of them. So, this makes it seem quite important. And, indeed, it's quite important. This begs the questions why a lot of companies don't ask more questions related to their customers instead of their competitors. Think of questions like, "In what way can we create as much value as possible for our customers?" Value is a process or service (or a part of it) that the customer pays or waits for. Before we can even attempt to answer the question, you need a clear view of who your customer is, to begin with. Are your customers predominantly female, between the ages 30-50, and do they like to read? These customer characteristics should be taken into consideration. To better frame your ideal customer, you can make a persona. A persona is a detailed description of a user of your product or service. Although personas are based on fictional characters, their specifications are based on real data gathered through talking with customers.

Furthermore, don't forget that customers aren't always external, you can have internal customers too, such as colleagues from a different department. It may seem common sense to focus on the customer to properly define value, but what is common sense isn't common practice. How often are things done "because they always went this way," "because the boss wants it" or "because we think the customer wants it"? Without a doubt, when you don't know what your customer wants, there's no way to manage anything adequately. Don't misinterpret customer value from customer demand. Although the difference may seem subtle, don't overlook it! Customer-centric companies understand extremely well what is of value to their (potential) customers. Thus, they "surprise" customers with a product that they had not yet asked for, but turned out to be of great value. For example, take Amazon, who started selling books and obsessively listens to customer's needs and desires. The customers wanted a wide range of books to pick and wanted these delivered as soon as possible. Amazon did exactly that by broadening the number of books available through its platform and building more fulfillment centers to ship the products to customers quicker. Thus, Amazon's revenue grew steadily, making it easier for them to "surprise" their (potential) customers with new gadgets, such as the Kindle and Alexa.

The second principle is about *identifying the value stream*; we need to determine where value is created. This can only be done after you have a clear image of the customer you want to serve and what they perceive as value and whatnot. Afterward, you take a look at your organization and see which activities add value to the customer and which do not. The second category, matters that don't add value to the customer, is called *waste* in the Lean methodology. A useful technique to make this evident is called *value stream mapping*. Thereby, you can map and detail processes and process steps from the trigger of the customer to the delivery of a service or product. Next, by taking a critical look at the created *value stream map*, we can determine for specific activities if they add value to the customer or not. So, we ask if *the customer value-added* or not. Lean distinguishes the activities that the customer doesn't value into two groups:

- Activities that add value to the business (Business Value Added): These are activities needed to keep the business going. You should minimize this as much as possible.

- Activities that add no value to the business (Non-Value Added): These activities are a complete waste of time and worthless. They aren't of value to the customer. First and foremost, and not to your business. What do we do with these? If possible, we want to eliminate them. Otherwise, we try to bring these worthless activities to a minimum.

To further illustrate this, say we run a marketing agency and want to give our clients more insights into market statistics. Delivering a short presentation with critical findings creates value for the clients. Now, before we could give this presentation, we had to be compliant with laws and regulations. Although being compliant isn't directly part of the value creation toward the client, it was necessary to do matters that create value, in this case, the presentation. However, if the presentation software we use crashes every couple of minutes, this is of no benefit to both the customer and the organization. Thus, this waste should be eliminated or reduced as much as possible.

The third principle *flow* is about ensuring a continuous flow in processes. This principle follows the previous step seamlessly because it focuses on removing all identified wastes from the process. By doing so, nothing remains but the value-adding activities. The next step is to adjust these activities to one another so that no congestion arises, and thus a natural flow is created. The ultimate Lean solution is the application of what professionals call a "one-piece flow," whereby (in contrast to batch production) valuable stocks are kept to a minimum, and errors or mistakes aren't passed on to the next process step. However, in practice, a more hybrid approach is, in many cases, the right solution. The purpose of creating this flow is to have the service or product "flow" to the customer according to her/his requirements without wasting time. This is only possible if the employees of an organization think from the chain of activities instead of seeing everything in separate boxes, i.e., looking at it in a vacuum. By making the lead time of the chain as short as possible, it's more the standard than the exception to have online orders reach customers the next day. Where "same-day delivery" is now something unique, this might be the standard sooner rather than later, because of the continuous improvement companies pursue in terms of shortening their lead time.

The fourth principle *pull* is all about doing what is needed when it's needed. With the previous steps, we've minimized the waste within the business process. Also, the processes are now far more focused on adding value to the customer, but something is missing, namely producing or delivering the service when the customer requests it. This limits unnecessary intermediate and final stocks. The outflow of products is the trigger for the organization to ensure new inflow. Of course, this principle doesn't mean that there isn't any stock at all. Take Walmart as an example. If the Walmart employees had to bake something like bread from scratch on the spot, that would take a lot of time, resulting in tremendous waste. Instead, *Pull* means that Walmart knows very precisely how much stock they need to keep. Thereby, they can replenish this stock based on previous sales data, *just-in-time*. According to Cambridge Dictionary, "a just-in-time system of manufacturing is based on preventing waste by producing only the amounts of goods(/products) needed at a particular time, and not paying to produce and store more goods than are needed."

The fifth principle *of perfection* is related to continuously learning and improving. With the rise of online communication channels, we're bombarded with messages and posts, wherever you go or don't go. Some posts proclaim that we shouldn't strive for perfection. Doing so would be mission impossible and would only result in distress, anxiety, and maybe a burnout. Why? Because perfection is something we'll never achieve with our fallibility. Therefore, this is not the idea behind this principle. The goal is to be a learning organization that does a little better day after day instead of large, usually uncertain, improvements now and then. A useful tool is organizing a weekly consultation or daily standup to structure the week or day's work. This makes daily improvement more ingrained in employees. Besides, it helps to ensure further improvements so that the chance for waste is eliminated or reduced to a minimum.

Lean Benefits and Waste

In the previous paragraph, we learned more about the various principles and that Lean is about creating more value for customers and eliminate or reduce all forms of waste. Lean distinguishes seven types of waste, often referred to with the acronym TIM WOOD(S), namely:

- **T**ransport. Moving products or materials (raw materials, documents, work in progress) between operations. Also, think of things like being physically separated from successive processes or an illogical layout of the workplace. Transport must be minimized because it takes time in which no value can be added and because products can be damaged during transport.

- **I**nventory. Keeping stocks of material and products, such as unread emails, pending customer cases, pending requests, or spare parts that are never used.

- **M**ovement. Any physical movement that does not add value to the process. Think of matters like unnecessary walking, lifting, turning, or reaching due to, for instance, the incorrect layout of the workplace or to search for documents.

- **W**aiting. Man or machine must wait for the completion of the previous process step. Waiting for authorization, starting a software program, instructions, and information are all examples.

- **O**verproduction. Producing more than necessary and doing more than requested by the customer. Also, things such as producing more information or documents than necessary, doing too much, starting too early or producing too quickly. This leads to stockpiling, yields nothing, costs time, space, and often requires management or maintenance.

- **O**verprocessing. Perform more process steps than necessary for the minimum for order handling. For instance, making the product "more beautiful" than strictly necessary, doing more than what the customer wants, such as picking up things outside the agreed standard service, doing things twice (saving files both physically and electronically), et cetera.

- **D**efects. This is the correction of mistakes made. For example, outages or errors, products or parts of services that do not meet the specifications of the customer, and incomplete or incorrect information.

- The seven wastes mentioned above are widely known in the Lean community. However, many organizations saw the need for an eight form of waste, namely, **S**kills. This is about the under-utilization of human potential due to the incomplete use of knowledge and creativity. Examples of this are over-qualified or under-qualified staff, too little use of employee capacities (an employee has a certain skill set that's untapped), and delegating tasks to employees with insufficient training.

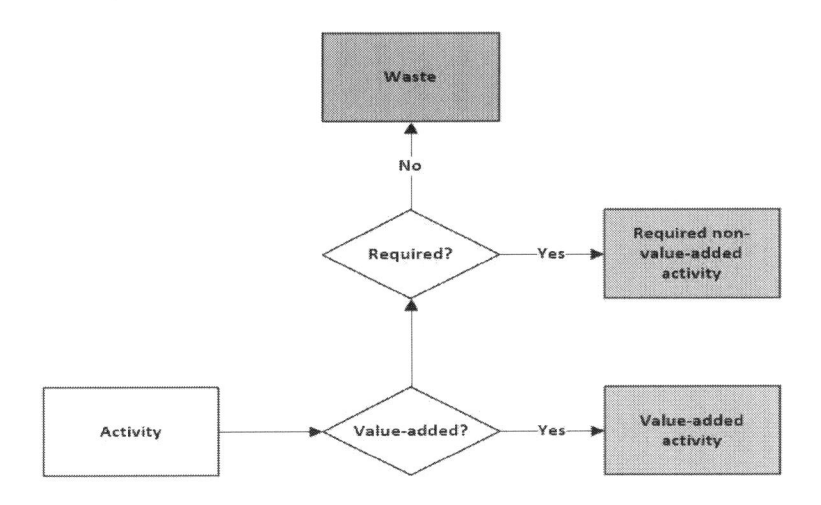

Image of a flowchart that starts with activity as a process. Then we check if it's value-added. If the answer is "Yes," this would be a value-added activity. If the answer is "No," it could be either a required non-value-added activity if it were necessary and a waste if it weren't necessary.

In later chapters, these wastes will be addressed further. But before we continue to the next chapters, it's good to provide you with a summary of the benefits we addressed in this chapter.

One of the greatest benefits that these Lean principles bring about is that they help you save a lot of time. This time can then be used to engage with customers to find new needs or wishes to create, keep, and deliver more value in a better fashion. Thus, the last principle of *perfection* brings us back to the first principle *value* to complete the circle. The principles mentioned above all bring about great benefits. In terms of the first principle *value*, a great benefit is that we always put the customer first. This has been the core of many successful businesses we hear about today, such as Amazon and Google. Putting customers first, has in and of itself a myriad of benefits, such as increasing customer satisfaction, turning more prospects into paying customers, and (usually) making more profit in the long term. Regarding the second principle *value stream*, a great benefit is that we create a vivid image of how processes run. With this knowledge, it's possible to identify bottlenecks, tackle these, whereby we create more efficient business processes. The third principle *flow* has many benefits too. For instance, if the processes run smoothly and don't inherit flaws from other processes, this will lead to reduced costs. Besides, the fourth principle *pull* shows us that doing more with less would benefit our business in the long run. Why? Because when we only do what is needed when it's needed, we drastically reduce the complexity of delivering our products and/or services. Finally, the fifth principle *of perfection*, has a lot of benefits as well, such as increased team morale. If you want to retain the great talent you've acquired, you need to challenge these professionals continually. If the work becomes boring or redundant, they'll easily leave you for a place where they can grow and are challenged more. When small improvements or "quick wins" are amplified in a talented team, it wants to strive to gain more and push new frontiers. In the end, this will create more value for the organization, but most importantly, more value for the customer.

Chapter 2. Six Sigma: Tools, Roles, and Concepts

In the previous chapter, we've addressed Lean. In this chapter, we'll take a look at Six Sigma. Six Sigma is the most effective methodology for problem-solving and improving the performance of business processes and the organization. All business, technical, or process challenges are far easier to overcome with the help of the Six Sigma methodology. The world's largest businesses have used Six Sigma to increase their combined profits by billions of dollars year after year, the past decade. Nowadays, in an increasing amount of organizations, competency of Six Sigma is necessary for any managerial position. Fortune 500 companies were first and foremost in adapting the principles of this methodology. In the past, it was always difficult for small and medium-sized businesses, public institutions, non-profit organizations, educational institutions, and even ambitious individuals to properly implement Six Sigma. This was mainly due to the scarcity of people experienced in this area. Most of these experts would be hired by large corporations who could pay the greatest sums. Fortunately, nowadays, things are different as the methods and tools of Six Sigma spread; more information about it was shared. Thus, it became easier to understand and easier to implement.

Simply put, Six Sigma is about applying a structured scientific method to improve an aspect of an organization, process, or person. It's about performing data collection and analysis, which helps you identify the best possible ways to meet your customers' needs and satisfy the organization's needs while minimizing wasted resources and maximizing profits. Six Sigma can be used anywhere. It's not only applicable in large and complex companies, but also the less complex and smaller businesses. Six Sigma is a rigorous and structured approach to problem-solving, for which you record data and apply statistical analyses to find out the real causes of the challenges that you encounter in production, service, or even transaction environments. That is why various chapters in this book describe and define various statistical tools of Six Sigma. But don't worry, these concepts will be explained in easy-to-understand language even if you don't know anything about statistics.

So, what's up with the name Six Sigma, you might think? You may know that sigma is the eighteenth letter of the Greek alphabet. But did you know that the term "six sigma" is used as it described a target of 3.4 defects per million opportunities? This small amount of defects in processes is near-perfect and considered world-class by organizations worldwide. Around 6 standard deviations around a central tendency covers this world-class result, namely 99,99966 percent accuracy (3.4 / 1.000.000 (x 100%) = 0.00034%, then subtract 100% from it). If the aforementioned standard deviations have a place within the specified customer requirements, we can confidently say that the process is Six Sigma competent. Through the years, this methodology has evolved to the point where it now contains several distinct aspects. These are listed below:

- First, as I just explained, Six Sigma performance is the statistical term for a process that produces fewer than 3.4 defects/mistakes per million defect options.

- Secondly, the methodology is about solving issues. Many organizations even see it as the most effective method for problem-solving when it comes to improving the performance of the business and organization.

• Third, a Six Sigma competent improvement occurs when the most critical outcomes of a business or work process are improved to a significant degree, usually by seventy percent or more.

• Fourth, an organization that strives to become Six Sigma oriented uses the methods and tools of Six Sigma to improve its performance by continuously reducing costs, increasing revenue, increasing customer satisfaction, expanding capacity and capabilities, reducing complexity, shorten the cycle time, and limit defects and errors.

• Fifth, proper execution is the prescribed roll-out of the Six Sigma methodology within an organization, with methods, roles, and procedures that are determined by generally accepted standards.

These aspects are closely related to the fundamental concepts within Six Sigma, with an order of importance; these are the following:

• The first key concept has *stable operations*. With this concept, Six Sigma organizations focus on ensuring foreseeable and unchanging processes to better the customer experience.

• The second key concept is called *critical to quality*. This is about the internal critical quality parameters that relate to the wishes and needs of the customer.

• The third key concept is the *process capability*. This is a measure of the ability of a process to produce consistent results. It's the ratio between the allowable distribution and the actual distribution of the results of a process. In short, it makes clear what your process can deliver.

- The fourth key concept is *a defect* and is any kind of unwanted result for your organization, but especially your customer. Because of the customer-centric view, it's generally defined as an error that means that at least one of the criteria for acceptance by your customers has not been met.

- The fifth key concept is *variation*. This is a statistical measure for the spread or variation. With this measurement, we get a better image of what the customer can sense, in particular, what she/he can see and feel.

- The sixth key concept is *design for Six Sigma* and is based on the premise that what we do as an organization is first cross-examined with what is best for the customer. Processes in your organization have to be designed in such a way that makes room for doing the same, i.e., meeting the wishes and desires of your target audience.

Knowing these critical concepts makes it easier for us to debunk the myths surrounding them. Just take people who say that Six Sigma only focuses on reducing defects. Well, this is far from the truth, because of the critical concepts, others don't deal directly with defects, such as *variation*. But why would we even care about defending Six Sigma, let alone apply it? Well, not using the concepts it contains causes many organizations to lose as much as twenty to thirty percent of their profits through mistakes that could've been prevented. This is a grave penalty for not working effectively and efficiently. Just imagine you had to throw thirty percent of your salary in the trash, every single month. This may seem outrageous, but this is exactly what most businesses do! Therefore, a good understanding of Six Sigma will benefit you, your organization, and your customers. So, let's take a step back and go to the time where it all began.

The Origins of Six Sigma

In the previous chapter, we took a look at Lean and how it originated. As you might've noticed, in this chapter, we addressed Six Sigma. Therefore, it seems logical to me to address its beginnings as well.

The start of Six Sigma came from the mathematician Carl Frederick Gauss. Gauss introduced the concept of the normal distribution, whereby the average spread of a measured value around a target value is expressed in the standard deviation. Six Sigma is a less accessible methodology than Lean due to the many statistical components that require more time. However, the method can be used in all kinds of industries. It's ultimately about customers and what they find important. Therefore, it's always good to first identify the wishes and needs of your customers through market research before you get started with Six Sigma. Six Sigma makes the probability that products and / or services are in accordance with what the customer wants as high as possible. Thus, customers are less likely to be disappointed. Although Six Sigma involves some statistics, by no means do you have to be a math genius to apply Six Sigma. Gauss did not come up with the exact method, but with his calculations, he laid the foundation for making variations in processes measurable.

Six Sigma is a management strategy that was originally developed by the American multinational communication company Motorola. Six Sigma, as a true management strategy, came to life at Motorola in the 1980s. The credits for Six Sigma as an improvement tool went to Motorola engineer Bill Smith when the company achieved a ten times reduction in product failure levels in a couple of years, something unheard of before. Unfortunately, Smith wasn't able to witness the hype around Six Sigma, because he passed away in 1993 in the Motorola cafeteria due to a heart attack. But this didn't stop the methodology from growing even further. Six Sigma matured further in large companies such as General Electric (GE), which managed to garner considerable savings by applying it. Soon after, General Electric became one of the most important users.

The same method is now used in pretty much any industry you could think of. Six Sigma has grown into a process management method that can be used in many different situations. Increasingly, Six Sigma is used by service companies, for example, by IT companies and transportation businesses. Six Sigma is particularly suitable for solving problems where the cause is not immediately apparent. A condition for this is that the quality of the relevant product and / or process must be measurable. Then the factors that influence the quality the most are improved. The most important approach of Six Sigma is quality management and the reduction of variation in production and business processes. Besides, discovering defects or errors and removing them is also important with Six Sigma. Therefore, the method consists of a collection of quality management methods. A Six Sigma project within an organization follows a certain sequence of steps in advance and also has financial objectives. These are five steps, which together form the acronym DMAIC, which stands for Define, Measure, Analyze, Improve, and Control.

Key Elements and Roles

Within the Six Sigma methodology, all matters go around three key elements, namely customers, processes, and employees. First, customers are the most important people to listen to. They give you more insights in terms of quality and other expectations. Customers are filled with expectations when they buy a product or service. Just think about it, if you order a physical product online, what things do you consider important? These are probably things like reliability, good prices, fast delivery, excellent service, et cetera. As an organization, you have to define and deliver these metrics to a high standard for the ultimate customer experience and, consequently, better customer satisfaction.

SecondF, an organization is nothing without its processes. Therefore, with Six Sigma, we spend a great deal of time defining processes, their metrics, and how we can adequately measure them because without measuring, we've no clue how matters are going. The processes aren't defined in a vacuum. Instead, these are represented by receiving customer feedback and looking at the processes from the customer's perspective.

Third, no process can run adequately without employees. If you wish to apply Six Sigma, all employees have to be on board and need to be involved. Of course, there's a clear learning curve involved in all this. To ease the learning, organizations can hold training sessions to instill key concepts further. Thereby, every employee should be able to use her/his skills to fulfill customer demands as well as possible. To properly get the ball rolling, the organization needs to take out some time to properly define roles and related objectives. Without doing the same, it's impossible to implement a robust Six Sigma methodology to satisfy your customer, because employees would, for instance, do tasks they are less suited for than a colleague.

Properly defining roles is crucial. All employees that are going to work with Six Sigma have a set of tasks to fulfill. In short, there are seven distinct sets of responsibilities. The first is the *leadership team*. This team is focused on getting all objectives defined properly for adequate Six Sigma processes. It can be compared to a Board in a corporate firm. Just as the Board defines a path, the same is done by the leadership team so that the employees meet the objectives they set. The leadership team has various responsibilities, such as: setting strict planning with deadlines of when specific work has to be completed, continuously explain how customer desires can be met even better, and supporting team members to grow their skills. The second is *the sponsor*. The so-called Six Sigma sponsor is knowledgeable about the process and is readily aiming for the most significant successes. Every Six Sigma project needs at least one sponsor to function properly. Usually, the sponsor is an executive of some sort and is readily focused on solving complex problems that pop up during the project. Besides, sponsors can be seen as owners too. They may not be the owners of the organization, but owners of processes instead. Improving and coordinating these processes is necessary for any good gains.

The third and fourth are the *implementation leader* or *champion* and *team leader*. The first has the responsibility of directing and motivating the employees who are part of the Six Sigma team. Besides, she/he supports the leadership team by communicating the completed work and possible obstacles the Six Sigma team faces. The second, team leader, has a more hands-on approach when dealing with the employees of the Six Sigma team. She/he communicates with the sponsor and makes sure the day to day tasks are fulfilled according to schedule. This schedule doesn't come falling from the sky. The schedule is created by the fifth role, namely the *coach*. The coach is a Six Sigma professional with the necessary certificates. She/he crafts the schedule defines tasks, and intermediates when conflicts arise. The sixth and seventh roles are as a *team member* and *process owner*. The team members are employees who work **in** the Six Sigma project itself instead of **on** it. The team member has deadlines to finish the stipulated work and communicates with other team members and the team leader. Finally, the process owner has the duty of ensuring that a process is fit for purpose. Usually, the responsibility of a process owner includes financing, design, change management, and continuous improvement of the process and associated measured values.

Process improvement is a challenge. If the right people with the right Six Sigma skills are involved, a significant and lasting change can be achieved. To define levels of expertise of the aforementioned organizational roles, every person involved in a Six Sigma project has a "belt color," as we see in martial arts. There are five important colored belts, namely:

> • First, the Yellow Belt. Employees with this belt understand the basic principles of Six Sigma. They pass on process

problems to colleagues with Green and Black Belts. Also, they participate in project teams and receive JIT (just-in-time) training.

• Second, the Orange Belt. These employees can apply the basic principles of Six Sigma practically. They have the possibility to manage small improvement projects.

• Third, the Green Belt. It starts and manages Six Sigma projects. Has Six Sigma expertise, but know fewer intricacies than people with Black Belts. Also, they give JIT (just-in-time) training to others.

• Fourth, the Black Belt. These employees report directly to a Master Black Belt. They have advanced expertise in Six Sigma. Additionally, they act as a coach, mentor, teacher, and project leader for project teams.

• Fifth, the Master Black Belt. This expert collaborates with leaders to identify gaps in projects and selects projects for improving processes. She/he provides coaching, mentorship, acts as a teacher, and leads projects independently. This expert is responsible for Six Sigma implementation and instilling a great culture in an organization.

Getting Started: The Steps and Tools

Before you can get started with Six Sigma, figure out if it's the right thing for your organization. For especially older, more static companies, change can be a tremendous challenge. Therefore, answer the below questions to get a better view of where your organization is standing at the moment:

• Would you say that the organization's strategic trajectory is clear?

• Would you say that the organization is efficient and effective when handling new situations?

• Has the organization evaluated the effectiveness of operation; how well are processes going?

- Can you make the needed investments (such as for training and consultation) and most probably lose in the short-term in order to win in the long-term?
- Are other aspects of the organization changing that are related to Six Sigma?

There are many more questions you could ask, but these are the most important ones, and answering these questions with Six Sigma experts should help you in making a readiness assessment to continue the effort or not.

If your organization decides to move forward, it has to draw out a path to achieve a Six Sigma organization. This is followed up by properly defining objectives, making these realistic and feasible, and gathering the necessary professionals. Afterward, the organization can immerse itself in creating the Six Sigma organization they desire by training its people and running a pilot Six Sigma project to test things out. The project could be assigned top-down or bottom-up. Both approaches have their up- and downsides. The first is heavily related to the strategy and customer needs, but the scope may be too wide. The latter has a small scope, but may not be very much related to the strategy. Six Sigma experts need to figure out a way between these two extremes to achieve the best results. As explained earlier, a Six Sigma project follows a couple of steps, namely Define, Measure, Analyze, Improve, and Control, which corresponds with the acronym DMAIC. Besides this method, there's another one with the acronym DMADV, which stands for Define, Measure, Analyze, Design, Verify. This book will expand on the first and most well-known method, DMAIC. The DMAIC cycle is a structured and proven method within Six Sigma to achieve better business results. Each phase within the DMAIC cycle contains useful tools and techniques that will guide you to achieve the defined organizational objectives.

DMAIC is a project-based and customer-centric approach to a problem, in which the causes and the solutions are still unknown.

Thus, a project tackled through the DMAIC phases has a definite start and end. Because the client's wishes are dynamic, various DMAIC projects follow each other up (changing customer wishes lead to new challenges), making room for a culture where life-long learning and continuous improvement are first and foremost. Let's take a quick look through all the fact-based phases before we delve into any intricacies.

- **Define Phase – What's the problem at hand?** The problem is defined in this phase. This is done by answering the following questions:
    - What's the problem?
    - What's the scope?
    - Which customer segment is affected by the problem?
    - What would the business case look like?
    - What is planning, and what are the critical phases in the process?

Attention is also given in this phase to the soft side of change. Think of things like how to select the ideal project team, who are the stakeholders, and how can I influence them. At the end of the Define Phase, both the project manager and the client have a robust view of what the problem is.

Tools and techniques that are often used in this phase include The Process Map, Voice of the Customer (VoC), and SIPOC (Supplier, Input, Process, Output, Customer).

- **Measure Phase – How big is the problem at hand?** If all is done correctly, after the first phase, we end up with a clear problem statement. Afterward, the relevant data is measured and collected in the Measure Phase. The following questions require consideration:
    - How do we make sure that the information we gather is correct?
    - What's the size of the problem?

o What information do I need, and what can I categorize?
o How specific must it be, and how will I collect the information?

The Measure Phase validates the business case that was illustrated in the Define Phase. With tools such as the Pareto Chart, Trend Chart, and Detailed Process Map, it's possible to sharpen the scope further. If the scope is well-defined, we continue with the next phase.

- **Analyze Phase – What are the root causes of the problem?** After the data has been collected and measured, it's analyzed in this phase. This includes attempting to recognize cause-effect relationships. Furthermore, the following answers are sought in this phase:
    o What is causing the problem?
    o What is the root cause of the problem?
    o Is further research needed? If so, what priorities do we set?

The Analysis phase approaches the possible causes from two sides, namely the data side and the process side. The process is analyzed to find out what regularly goes wrong, what adds value, and what doesn't. The data is used to substantiate the assumptions so that it's clear to everyone that the defined causes are indeed the root causes. To achieve this, we can use various tools and techniques such as The Histogram, 5-Why Analysis, and Regression Analysis.

- **Improve Phase – What countermeasures do we implement?** Once the root causes are known, it is time to figure out improvements, test them, and implement them. For this phase, consider the following:
    o Know that the focal point remains on the best solutions. These solutions have a high probability of success and cost the least amount of resources.

o Ensure that the solutions are tested in a pilot. If the solution is tested successfully, it can be implemented in the real organization.

The most challenging aspect of this phase is the possible resistance the project leader will face for implementing the countermeasures. Therefore, in this phase, extra attention is given to resistance tools and techniques like brainstorming, Poka Yoke, and Deployment Flowchart.

- **Control Phase – How do we safeguard the countermeasures?** This phase is dedicated to safeguarding to prevent failure. The English verb "to control" in this context means "to manage." Related questions in this phase are:
  o How can I ensure adequate results and prevent failure?
  o How are we going to respond to signs of failure?
  o How can I visualize developments and identify trends early?

A successful project stands or falls by an excellent guarantee. If insufficient attention is paid to this, you may win in the short term, but will be very disappointed in the long term, because of the lack of competent assurance mechanisms.

In the coming chapters, all these phases and their essential tools and techniques will be addressed in more detail.

Project objective
- Problem statement
- Impact on customer

Measure system
- Measurements
- Measurement plan

Guarantee
- Evaluation
- Process assurance

Improvement idea
- Solution and testing
- Implementation

Root cause
- Analysis
- Validation

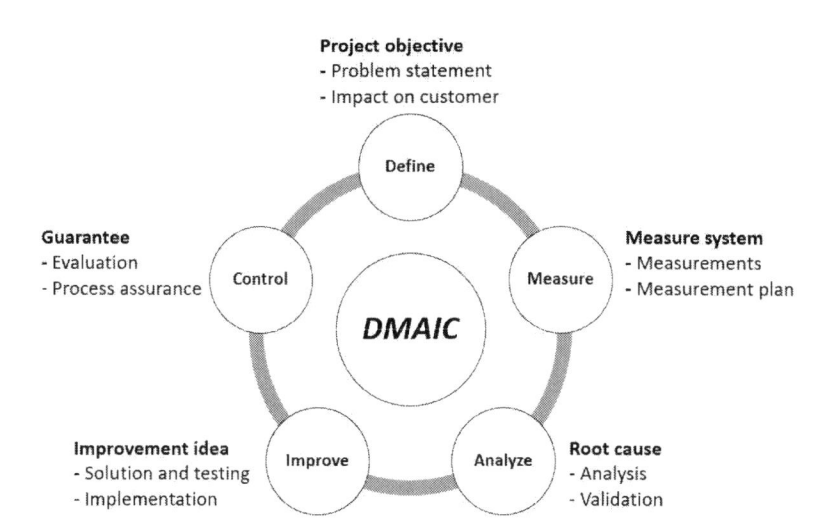

Image of the DMAIC Cycle. Define – Project objective, problem statement, impact on the customer: measure – Measure system, measurements, measurement plan. Analyze – Root cause, analysis, validation. Improve – Improvement idea, solution, and testing, implementation: control – Guarantee, evaluation, process assurance.

Chapter 3. Lean + Six Sigma = Lean Six Sigma

The first chapters were dedicated to two methodologies, namely Lean and Six Sigma. But what happens if we combine both these great methodologies? Well, Lean + Six Sigma = Lean Six Sigma. This is a new methodology that combines all benefits found within these two approaches of managing customers, projects, and processes. Lean Six Sigma is a globally proven methodology to attain sustainable forms of improvement of processes and organizations. The methodology proffers an approach by which organizations can achieve concrete results with continuous improvement in a structured manner (Define, Measure, Analysis, Improve and Control, or DMAIC). The focal point is on what customers find important and to realize this in processes. No more and no less. With this approach, costs are reduced, customer satisfaction is increased, and the lead time is shortened. The strength of this approach is that we make use of existing knowledge and experience of people within the processes. From the Lean Six Sigma methodology, both the process and the quality improve. Lean strives for more flow and value generation. Six Sigma strives for stable and effective processes. In combination, they reinforce each other and are entirely complimentary. Lean is for waste elimination and Six

Sigma to reduce variation. Do you remember more about what you read about these methodologies? No worries if you don't remember much. Here are the key points to reinforce the ideas in your mind.

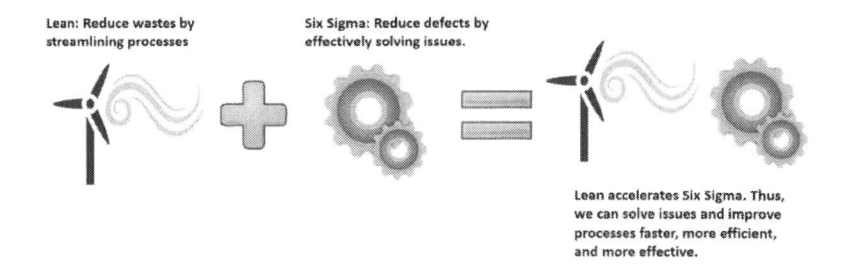

Image of a windmill with the text: "Lean: Reduce wastes by streamlining processes," an addition sign, an image if gear with the text: "Six Sigma: Reduce defects by effectively solving issues." Combine these images, and we get the following text: "Lean accelerates Six Sigma. Thus, we can solve issues and improve processes faster, more efficient, and more effective."

The basics of Lean can be deduced from the Toyota Production Systems (TPS). The Japanese inventor Taiichi Ohno is the founder of this system and based it on developments from other inventors such as Henry Ford and methodologies such as Business Process Redesign.

- Lean starts by determining the added value for the customer, a product, or service that meets certain conditions or specifications.

- The process that delivers this value is then mapped the so-called value flow.

- By carrying out this exercise, it becomes clear where waste is located, and an improved flow process can be made. These wastes become apparent through the collection of data around them, and together with the employees working in and on the process, it gets better step by step.

- Then, you look at how you can set up the process so that it starts when the customer requests it and then delivers precisely on time when the customer wants it.

- People in the organization see where improvements can be made and work on these. In this way, an organization is created in which there is a process-wide approach, and everything is focused on doing what the customer demands in a smart, effective, and efficient way.

Regarding Six Sigma, it was developed around the same time as Lean - and is more data-driven. The methodology has been developed by Motorola and has been widely and successfully applied by General Electric and many other organizations, small and big. The structured project approach Define, Measure, Analysis, Improve, and Control (DMAIC) ensures that the root cause of a problem is first found before a solution is implemented. By applying the DMAIC Model, you use the capacities of your employees within the organization structure efficiently and intelligently to create as much value as possible for the customer. A variant of this model is the DMADV: Define, Measure, Analyze, Design, and Verify or DFSS: Design For Six Sigma. By determining and measuring the critical indicators of the company, it becomes clear where improvements need to be made. The management receives targets and tools for this through Six Sigma, thus creating a structured improvement model. As with Lean, this is based on the added value for the customer.

There are many benefits to implementing both methods in the form of one way, i.e., Lean Six Sigma. Think of the following benefits:

- Lean Six Sigma increases the productivity and profit of your organization by improving the efficiency of your processes. Efficient processes result in products and services that are completed faster, without compromising on quality.

- Furthermore, Lean Six Sigma reduces costs for your organization by eliminating all activities that contribute nothing to the creation of products or services. By removing all these unnecessary activities in your processes, you also reduce complexity and improve communication in both your company and with your customers.

• Besides the application for specific improvement projects, Lean Six Sigma is very important for managing processes organization-wide. It provides a framework for reducing product defects, improving returns, limiting warranty costs, and improving delivery quality.

• Lean Six Sigma helps you change your organizational culture so that people have a similar understanding of doing business. This helps to guarantee the quality of your products and services. Your company changes from a reactive to a proactive organization that focuses on continuously satisfying customers, be they internal or external.'

Lean is focused on fast and efficient business processes, whereby quality assurance often turns out to be an issue. It does not matter how many forms have been filled or documents created if the stored information does not meet the requirements. Quickly completing activities without quality assurance leads to an environment that is error-prone, resulting in redoing the work. This is where Six Sigma becomes an essential part of process management. The Six Sigma methodology is a quality tool that emphasizes the reduction of errors in a process. It focuses on identifying variation in those factors that influence the result and looks for the real cause of errors with a Root Cause Analysis. Nowadays, Six Sigma plays a vital role in the leadership of an organization. Large-scale implementation can help a company achieve real and measurable results. From a distance, the two methods almost seem to be opposites. However, many practitioners argue that the best approach to achieving an effective and efficient business structure is to implement both Lean and Six Sigma. Besides, if a company opts for this combined implementation by two specialist practitioners, this can lead to a corporate culture where *thinkers* (Six Sigma) and *doers* (Lean) combine forces.

Both forces need to receive the correct training, so let's look at the possibilities for training and certifications in the next chapter.

Chapter 4. Training and Certifications: What Do I Need?

There are various ways to garner the necessary skills to conduct projects with Lean Six Sigma. Just like the Lean Six Sigma efforts themselves, the training programs vary from organization to organization. However, in practice, we tend to see various similar elements when training employees.

First, when an organization is entirely oblivious to Lean Six Sigma, it's best to start with an awareness course, i.e., White Belt training. The White Belt training gives a great overview and teaches the basics of Lean Six Sigma in a few hours. The aim is to introduce people to the language and concepts of Lean Six Sigma. Working on a project is usually not a condition.

Second, the organization can take up a more in-depth introductory course in the Yellow or Green Belt training programs. This next-level includes a direction in which people can practice using the improvement methods and tools. In larger organizations, this training can take up to one to two weeks. The Yellow Belt is an intermediate step between the White Belt and the Green Belt. The training of a

Yellow Belt is usually a bit shorter than that of a Green Belt. If employees want to become a Green Belt, they are usually expected to be able to manage projects. If someone wants to become a Yellow Belt, she/he only needs to participate in a project but doesn't have to lead it.

Third, the organization can focus on training a few professionals to become especially skilled in all facets of Lean Six Sigma, with the Black Belt Training. The Black Belts form the core of the Lean Six Sigma support structure. In some companies, they manage projects; in other companies, they work as coaches on different projects simultaneously. Fourth, for an elite group of professionals, it's possible to go even further, with a Master Black Belt training. This training consists of a handful of refined Lean Six Sigma tools. Not every Black Belt needs to learn these techniques because they are not as common as the general Lean Six Sigma instruments. These skills are often offered as part of the training for a Master Black Belt or in specialized in-depth courses for Black Belts. A full Black Belt training usually takes four to six weeks, one of which focuses on leadership training. Typically, the Black Belts follow one week of training, then they work on a project for a couple of weeks, return for training a couple of days and get back in the trenches. It's a very hands-on approach to learning.

The training consists of Lean and Six Sigma methods, "complexity reduction" methods, as well as project management and leadership skills. Each participant receives one to five days of support from a Master Black Belt with experience in implementing improvements and managing projects from different sizes and organizations. In terms of forms, training can be followed through online courses or can be conducted in person, for instance, by hiring a Master Black Belt consultant in house. With many courses available today, participants have access to both electronic and printed training materials, case studies, and other resources.

For each of the belts mentioned above, it's possible to get certified. For most organizations, it's not needed to acquire a certification.

Having the basic knowledge of Lean Six Sigma can be acquired through other sources too, such as the book you're reading now. Properly going through this book will give you more than enough knowledge to get going with Lean Six Sigma without any additional training. However, if you would like to acquire a certification (perhaps your ideal job requires this), pay attention to a few things. When you attempt to get certified, always make sure the training organization is appropriately recognized and legitimate. For instance, The International Association for Six Sigma Certification (IASSC) delivers training programs and accredited certifications upon completion of exams.

Chapter 5. The Lean Six Sigma Process: DMAIC VS DMADV

Now that we have a good overview of Lean, Six Sigma, and Lean Six Sigma, it's time to focus on the practical side of Lean Six Sigma. In this chapter, we take a closer look at the Lean Six Sigma process and how you can implement the same.

As explained previously in this book, there are two critical methodologies for implementing Lean Six Sigma in your organization, namely DMAIC and DMADV. In this chapter, you'll learn more about both these natural processes and their benefits, so that you can decide the process you'd like to follow.

DMAIC

So, what is DMAIC about? As explained in a previous chapter, it consists of five phases that seamlessly merge into a cycle process and is an acronym for Define, Measure, Analyze, Improve, and Control. Let's take a look at each phase:

- **Define.** The problem is defined in this first phase. In addition, it is important to recognize and define the following components:

- What is the goal, and what are the related business processes?
- Who are the customers?
- What are the critical phases in the process?

- **Measure.** In this phase, the most important aspects of the current process are measured, and relevant data is collected. The following points are essential:
  - Analysis of input and output.
  - Definition of the measurement plan.
  - Testing of the measuring system.

- **Analysis.** Subsequently, the investigated data is analyzed. We look at the different cause-effect relationships. The root cause of defects and errors is sought out. Essential tools and techniques are used to:
  - Detect gaps between the current performance and the desired performance.
  - Identify the input and the output.
  - Prioritizing potential opportunities.

- **Improve.** The current process is improved in this phase with the help of techniques and creative solutions. Brainstorming sessions can be a useful tool. Other apparent solutions are:
  - Innovative ideas.
  - Focus on the simplest and most comfortable solutions.
  - Prepare a detailed implementation plan.
  - Implementation of improvements via, for example, a quality circle.
  - Finding out errors and causes, using an Ishikawa diagram.

- **Control.** This phase is not just about control, and it's more about supervision.
  - Control takes care that any possible defects in the future are prevented as much as possible. Supervision

leads to lasting improvement and therefore guarantees long-term success.

In addition to the standard DMAIC model, it's wise to also implement this step-by-step plan in other business processes. By sharing experience and newly acquired knowledge with other departments, changes can be made more efficiently throughout the entire organization. It's essential that employees have a good understanding of the usefulness of the working method of the DMAIC model, that they can discuss this well with each other, and are willing to share experiences. The DMAIC model was initially linked to Six Sigma and is intended to improve the quality of the production (output) of a process. This is done by identifying and removing the causes of errors. However, the DMAIC model is not exclusive to Six Sigma and can, therefore, be used to improve processes at other organizations.

Furthermore, The DMAIC model is an application of the PDCA cycle of William Edwards Deming. Where the DMAIC model is based more on a project-based approach to the problem, the PDCA cycle is both widely applicable and applicable to one project. The DMAIC model analyzes the root cause of the problem, while the PDCA cycle focuses attention on the whole and also uncovers other causes. Besides, the DMAIC model is based on framework thinking within, for example, product group, customer group, or service department. The strength of the model lies in tackling and optimizing root causes of problems in a process. However, the model is less applicable for a completely creative change in which the course within an organization is completely reversed. When going through the steps in the DMAIC model, there should not be any overlap. The best results are achieved with a planned team approach.

To further illustrate this, take the steel manufacturer NewSteel I just came up with. NewSteel has 20 locations nationwide and is specialized in car frames. Until now, they have plenty of stock of various types of frames in every branch. This allows them to help the customer quickly. This subsequently results in satisfaction and good word-of-mouth advertising. Now it appears that last year the costs of the steel manufacturer have increased rapidly. Therefore, the management decides to enter into a dialogue with all 20 location leaders and find out what could be the cause, by using DMAIC.

- **Define.** The costs have gone up considerably and include personnel costs, renting the 20 properties and inventory.
- **Measure.** The costs are compared with the norm of the previous year, and it appears that the costs are 20% higher, without more turnover being generated.
- **Analysis.** In consultation with all 20 supervisors, the largest cost item was examined. This reveals various factors. On average, it appears that the inventory costs at all locations were 15% higher than the year before, which is partly due to too little variety, which means that "older" models of frames remain in the warehouse longer.
- **Improve.** Together with the 20 branch location leaders, the management is looking for possible improvements. For example, the proposal is to work with a central warehouse, from which the locations will be supplied two to three times a week. Another possibility is to only stock standard frames and to store the more specific types of frames in the central warehouse. After a discussion with the NewSteel supervisors, they selected the first solution.
- **Control.** After a test period, an evaluation takes place with all supervisors and the management. It appears that NewSteel often has to sell "no" to the customers or that customers can only be helped with a delay of several days. After this evaluation, they considered the second solution they had at their disposal and decided to have the most commonly used frames in stock per location. Also, they decided to deliver the remaining frames from the central warehouse. Of course, a test period and evaluation will take place after this choice, and the costs are also passed on.

DMADV

Now, what is DMADV about? DMADV stands for:

- **D**efine
- **M**easure
- **A**nalyze
- **D**esign
- **V**erify

It is a quality method used for designing new processes to ensure that, when the end-product goes to the customer, it is delivered correctly. The method's purpose is to create the highest quality products, keeping the end-customer and their needs in mind throughout each of the five phases listed above.

With Six Sigma, the efforts are related to process improvement within quality management. Existing products and/or services are improved through analytical techniques and statistics. Six Sigma focuses on reducing variation in issues that are perceived as Critical to Quality (CTQ) by the customer. These CTQs are very important and crucial for quality; it concerns the internal critical quality parameters that relate to the wishes and needs of the customer. CTQs are, therefore, quality properties of the process or service that meet what the customer considers important.

The DMADV model revolves around the development of a new service, product, or process. The DMADV model is especially useful when implementing new strategies and initiatives. Consider the following phases:

- **Define.** This first phase of the DMADV model is about identifying the purpose of the project, process, or service. Not only seen from the perspective of the organization but also from the perspective of stakeholders, including internal and external customers. It must be clearly defined which guidelines are essential for the development of a product or service, whether there are potential risks and what the production planning is. During this first phase, the project leader finds out the customer's most important requirements as far as the service or product being developed is concerned. How? By using customer information relevant to the project and feedback from the customer. As an example, look at a company called GardenJoy. They make garden furniture, and, in the first phase, they may opt to focus solely on garden benches constructed of wood. Because of the information they obtained from the customer earlier, they know that the customer wants fair trade wood used. As well, the customer has said the benches must be big enough for a minimum of two people, the headrest and backrest must be comfortable, and the coating must be environmentally friendly; that way, they can leave the furniture outdoors all year round. In this first phase, the manufacturer will also consider whether a bench such as this would be profitable to develop.

- **Measure.** The second phase is focused on the collection/recording of data relevant to the measures (CTQ) that the first phase identified. This data is critical as it drives the entire process. In the DMADV Measurement phase, CTQs do not exist. Why? Because their product hasn't been manufactured so we can't even begin to examine the production process. Instead, we look at exactly what the customer requires; these factors then get linked to the quality, and the CTQs are then created. If every design component is assigned a value, it follows that a practical approach should be created for getting the production process off the ground.

Now is when it is critical to determine the production parts important to all the stakeholders. Ultimately, the customers' needs will be turned into clear objectives, ensuring a product that stands apart from its competitors is created. GardenJoy now takes the important customer requirements and links to them to the CTQs. If fair trade wood cannot be procured, the production will not go ahead. That also applies to the coating

– if it isn't environmentally friendly, it can't happen. And lastly, the design – if the requirements of at least two spaces, a comfortable headrest and backrest cannot be met, the production will not begin. In this phase, the manufacturer will also look at the costs of production, design, and materials to see if they are more or less than the final price.

- **Analyze.** This DMADV phase goes together with the previous phase; at the time the data is collected, the project team must also analyze it. With this, we get the best basis for improvement measurements during production. It is also in this phase that the alternative designs are developed, and all the different requirement combinations are looked at. Here, an estimate is produced of the total costs of the entire design process and, once all the design alternatives have been considered, a rough design I the product is made, meeting the CTQs as near as possible.

Here, GardenJoy examines all the potential importers for fair trade timber, looking at where the wood originates so the information can go in the sales report. They look at the different coatings, working out the pros and cons of the environmentally friendly ones, as well as looking at the quality.

This is one of the most time-consuming phases because, on top of all that, the team must also analyze all the different designs. Here, the manufacturer would benefit from having a deadline in place; otherwise, the costs could spiral out of control.

- **Design.** In this phase, the product or service design is drawn up to the customer's requirements. The project team takes all the previous data and ensures that the product or service is suitable and that all the adjustments that can be made are. The design is incredibly detailed and of high quality, working as a prototype and includes consideration for the production process. The idea is not just coming up with the process to manufacture the right goods; it must also be an efficient process, in all areas. GardenJoy has considered their analysis and made their decisions – they have a fair-trade wood supplier; they know which of the environmentally friendly coatings to use, the design includes a safe, comfortable adjustment, and the right seat back and headrests. With manufacturing, one of the critical considerations must be how the machinery is laid out to ensure the process is fast, safe, and efficient in that the highest possible number of benches may leave the factory.

- **Verify.** This may be the final phase, but it certainly isn't where the process ends. For the quality to be guaranteed, the product must be continually checked and adjusted where needed. The design is completed, and the product may now be sold. During the Verify phase, the customer provides the team with feedback on the product and their experience; this is used to tweak the product to ensure it meets the exact requirements of the customer. The team also forms extra CTQ measures, so the feedback can be followed once the product has been delivered.

For GardenJoy, they must use this phase to see what the customer thinks. Is the fair trade wood good enough? Is the bench comfortable? Easy to use? Is the coating, right? And so on. If, after a month or two, they start to see negative reviews about the product, they know that something needs to be changed, and the data collected in previous phases is used to make those changes.

DMAIC VS DMADV

DMADV and DMAIC are similar; both are easily utilized with Lean Six Sigma and are identical in some ways. One of the most critical of these similarities is that they both aim to improve business processes, ensuring they are both effective and efficient. While the first three letters in each acronym are the same, they are not interchangeable and are, in fact, each used for a different purpose.

The primary difference is in the final two steps of each model; DMAIC focuses on improvement and control of the existing process while DMADV focuses on the development of new service/products. As such, it is in the second model, where the customer requirements are emphasized. When a new service or product is developed, it must include everything the customer requires- and at the best of quality. Product designs meeting the criteria are known as DfSS, or "Designed for Six Sigma." More often than not, there may be a third criterion to DMADV – logistically, the production process has to be highly efficient. Then, we can talk in terms of DfLSS – "Designed for Lean Six Sigma."

Additionally, DMAIC carries out interim checks, improving the process where needed and reducing or eliminating defects. With DMADV, the focus is on the development of suitable models that fully meet the requirements of the customer, as we saw in the GardenJoy example. While the DMAIC cycle focuses on eliminating sub-standard quality, the DMADV cycle focuses on developing new quality features. This difference is amplified, especially in the Measure phase. At DMAIC, Critical to Quality (CTQ) parameters are measured that are most frequently outside the specifications. Afterward, the causes are identified and noted.

On the contrary, with DMADV, there are no CTQs at all in the measure phase. After all, there is no new product yet, let alone a production process! In this case, the measure phase includes determining what the customer considers important for a new product. The goal is to arrive at a Design Scorecard. This Scorecard contains the characteristics that the product must have to be useful in the market. In short, the Measure phase ultimately yields new CTQs. The Scorecard is used in the remainder of the product development process to check whether the design has these features.

All in all, both approaches DMAIC and DMADV can be used in specific contexts. It's up to you as a project manager to select the corresponding method. Are you willing to improve an existing business process? Then DMAIC is the way to go. Are you ready to (re)design a new product or service? Then DMADV is the way to go. Because this book is more about dealing with existing processes with the help of Lean Six Sigma, we'll delve deeper into the DMAIC approach. Let's get this going!

Chapter 6. Define: Process Mapping and Customer Voice

Every larger organization knows the situation that a problem seems to be solved, and yet it reappears day after day. The whole organization works hard, sometimes for months on end, to fix the issue. Now it will finally be solved, is what goes through the minds, but again things go wrong. An organization that works with Lean Six Sigma cannot afford such failures and therefore uses a more robust and dynamic solution method, namely DMAIC. DMAIC has proven to be one of the most effective problem-solving methods. Teams are forced to work with data. Data is important to achieve goals, such as finding the root cause of a problem and what processes the problem is related to. In this chapter, we take a look at the first phase, namely the *Define Phase*.

What is the Define Phase?

The Define Phase is the phase where the problem is identified. The purpose of this first phase in the DMAIC process is that all members of the team and the sponsors agree on what the project entails. There are a few activities to undertake in this phase:

- Search and gather data about customers for customers.
- Check the existing data about the process and/or the problem.
- Ensure that the entire team is involved in the project charter, a key element in planning a Lean Six Sigma project.
- You outline the improvement process in the outline. This type of process overview is a widely used improvement tool in DMAIC. In the "Define phase" of the DMAIC process, they are used to indicate the limits of the project.
- Prepare a plan and write guidelines for your team.

Besides these activities, consider these matters that follow, below.

These activities help you to clear up misunderstandings about the how and why of the project. Everyone employee on the team must be on the same line. A team often gets stuck as soon as there is disagreement about which data has to be collected or which solution is the best. Such a situation can occur if employees do not realize that they have different views on the project and the objective.

Talk to your customers and study the data that you have available. This is essential because then you know for sure whether the solutions also have a chance of success. It helps you to refine the goals of the project further. In exceptional cases, a project can be canceled if the data shows that it is not useful to continue working.

Come to an agreement with the management about a realistic scope of the project. If your team believes that the project is too large or too small, consult with management about possible changes. For example, you can add resources or adjust the deadline.

Make clear agreements about the standards for measuring success. A common mistake is that teams and stakeholders have not determined in advance when the project can be called successful. For example, a team is satisfied because it sees the number of errors decrease, but the only thing the manager wanted was an increase in sales. A team needs to know exactly what a sponsor pays attention to when assessing the result.

The All-Star LSS Team is formed and gathered to achieve worthwhile success. A team can consist of employees who already work together daily, but also of people from different departments. So it may be that the team members do not know each other. At this stage of the process, they have the opportunity to get to know each other.

Process Mapping: SIPOC

A core principle of Lean Six Sigma is that a defect can be anything that makes a customer unsatisfied, whether this is long delivery time, variation in the delivery time, poor quality, or high costs. To address these issues, be the first to look at the process in your company. To what extent does your company meet the specific wishes of the customer? What influences a process, and who has to deal with it? A SIPOC provides insight into this. It maps out the process that you want to improve and helps you determine the scope. It's an acronym for:

- **S**upplier
- **I**nput
- **P**rocess
- **O**utput
- **C**ustomer

The supplier provides the input for a process (step), and the output goes to the customer. The output must meet or exceed customer requirements.

There are many benefits to working with SIPOC. A SIPOC model can provide some insight into possible areas for improvement that can be explored further in a later stage. Above all, the purpose of a SIPOC is to define the boundaries of a project (scoping) by displaying a clear start and endpoint and ensuring that the process is divided into 5 to 7 well-arranged process steps. Another benefit is that this tool adds visualization to the conversation about the process. This helps the team members to gain a lot of value that would otherwise be lost in translation by only speaking without visualizing. It structures the conversation you are having and facilitates the discussion about the process. In addition, you acquire a good view of the customer and other stakeholders. Ignoring important stakeholders damages the relationship, and it will cost you time and energy later in the project to restore this. And above all: because it teaches you as a project manager to quickly come to the essence of the problem and process!

How does this look like in practice? In the first conversations with the stakeholders, you ask the customer to tell more about the process that needs improvement. You do this based on the following steps:

- Draw the structure of the SIPOC on a flip chart or whiteboard and grab some post-its with you.
- Determine the input? This can, for example, involve a telephone call from a customer, a delivery of raw materials, or a complaint. It is often a noun.
- Who supplies this input? (this is the Supplier!)
- What is the first thing that happens with the input? (that will be the first step in SIPOC, registration here)

You then work out the successive steps until you reach a concrete result:

- What is the result (which output)?
- Who receives this output?
- The recipient of the output is the customer of our process. This "customer" sets requirements for the output, for example, when it comes to quality (good solution in terms of content), but also to timeliness (within which period?), or to the method of delivery (personal, e-mail, social media, et cetera).

- And then, what requirements does 'the process' make of the input? This is very important because if your inputs are bad, your outputs will be bad too. Remember garbage in, garbage out! Therefore, the input must also be of good quality to be properly processed.

With the SIPOC model, we can see which department delivers what part of the production and what the result is. It is also much clearer who the customer is. Below, you can see the SIPOC model elements are, explained using a real-world example:

- **Supplier** – the person/people who contribute to the entire process; they may be internal to the organization or external. We answer the question of what is required and used for the process in the next point, but this one is about who supplies the input. This is the party that supplies everything needed for the process, be it raw materials, information, skills, or knowledge. Going back to GardenJoy, they need several different suppliers to produce a garden bench. They need materials, designers, software suppliers, specialists, and employees. All of those are suppliers because they each supply something for the manufacturing process – raw materials, drawing and design software, designers, specialist employees, and so on.
- **Input** – now they have the suppliers, it is much easier to work out what is delivered, and that means you can work out what the inputs are. The input section is all about the service, materials, and information required by one or more of the process steps. What is required for a step? GardenJoy requires raw materials in the form of wood; they need glue, nails, and handles, and the software must be applied correctly. On top of that, they need the right tools and machines to do the job.
- **Process** – this is a series of steps that make sure the input gets converted into the right output – the service or product. Typically, a manual, documents, or set of instructions is used

to record this; provided each step is described properly, the output will be correctly delivered with few to no deviations. Back to GardenJoy, by doing this, they can ensure that their product leaves the manufacturing plant in bulk, barely differing from one another and all of the exact same quality. GardenJoy has a simple process – draw up the design, bring in the materials, construct the garden bench, package, and ship it to the relevant customer.

- **Output** – this is all about the end result, be it a service, a product, or information going to the customer. It covers what the steps deliver and what the customer requires for the end-product. Another part of the output that must be considered is waste – for GardenJoy, think of the bench as the end product and things like wood splinters as wastes.
- **Customer: Who are the customers?** This is any party that receives output from the process. This can be organizations, people, departments, but also *systems*. Who is the output for? The person or thing that receives the product or service is the end-user or customer. That may be the actual customer who purchases furniture from the factory, but it can also be the employee who wants to buy furniture for her/his department. The customer requirements are important for the entire process. If the output does not meet the customer requirements, the quality cannot be guaranteed. GardenJoy focuses on selling Business to Consumers (B2C) and Business to Business (B2B).

The above shows that we only mention first-line suppliers and customers within the SIPOC model. This means that other stakeholders, such as banks, subsidy providers, or licensing authorities, cannot be found in the SIPOC model. Besides, it's not common for the quality properties and/or specifications to be displayed. Furthermore, it's crucial to craft a SIPOC diagram with a group of employees, allowing all stakeholders to contribute to the process. Doing this in a vacuum will result in much confusion and brings the process in danger. Employees can bring their insight to the table and determine their role within the process. A SIPOC model creates mutual understanding and a shared view of the process and the reason behind it, and this leads to a high degree of involvement. This prevents uncertainty in the later phase of the process.

Now that you know what the SIPOC diagram is all about, let's fill it in! The steps you should take to fill in the SIPOC diagram are the following:

- **First, print or draw a SIPOC diagram and give it a clear title.** If you want to draw it out, make sure there's enough room to write. Now that you have a SIPOC diagram in front of you, ensure you have a clear name for the process you want to map out. Put this above the diagram as a title.

- **Second, accurately define the starting and ending points of the process you want to improve.** Often, these can be found on the team charter in the "scope" section.

- **Third, jot down the remaining parts of the process, around 4-8 main steps.** Don't do more than eight steps, or the process becomes long-winded. Don't do less than four, or the process becomes poorly-defined. These are the raw and most critical steps in the process. They don't include any decision points or feedback loops.

- **Fourth, write down the most critical outputs of the process you're tackling.** In practice, we see that the list of outputs includes around three or four main items.

- **Fifth, adequately define who is going to receive these outputs: who are the customers?** Again, these customers can be internal or external, and it can be a person, organization, or even a system or machine.

- **Sixth, jot down the inputs needed for the process.** Like the outputs don't go overboard in mentioning every intricacy. We're not interested in knowing that a laptop and stapler was used to print a document and staple it.

- **Seventh, for the final step, define and list the actors who supply the inputs to the process.** Do you make use of manufacturers? Are there specific professionals in your organization needed to make the process work? These questions help you define who supplies the inputs to the process.

Voice of the Customer: Identifying and Understanding Customers

The goal with the Voice of the Customer (VOC) is collecting and recording customer requirements/expectations for the process you want to tackle. We apply the VOC when problems/assignments are linked to the customer, and this is usually the case. Mapping requirements is essential: if you don't know what the requirements are, how can you know if a process is causing problems? You can't. Therefore, take the following steps:

- Discuss and align with the owner of the assignment or project. What does she/he want to achieve? The owner of the project can be the team leader, but also the CEO, depending on the size and culture of the organization.
- Interview employees who readily talk with customers, such as customer service representatives.
- Approach customers and talk to them. For the most part, having a relaxed conversation is advised, but also make room for some more in-depth interviews with a selected group of customers. Besides, it's possible to garner more information with a survey. There are many websites whereby you can do a survey and have it promoted to your ideal customers, such as SurveyMonkey.
- Record any customer requirements/expectations that come up. You can use a tool such as a Kano model for this. Afterward, possibly translate this into indicators and specific standards.

After having a clear image who the different customers or customer groups are, you can start defining the VOC using these points of advice:

- Before using any online questionnaire, try to collect information directly from the mouths of the customers, based on open questions. Use techniques like the "5x why?" Method (explained in a later chapter) to get to the heart of the wish. Clients reason from their existing frame of reference and are not always aware of the possibilities that a product or service has to offer.
- Organize a Customer Journey Mapping session, in which you go through your process from the customer's perspective. When do we have contact with the customer? And how does the customer experience our process?
- A mistake that many make is to think that mapping the Voice of the Customer is a *one-time* process; both customers - and their wishes - are continually changing because they change over time. This process is accelerated by the technological developments that we see all around us. Because the Voice of the Customer is continuously changing, as an organization, you will have to stay in constant contact with your customers.

How we make the VoC concrete is by attaching critical quality characteristics to the wishes of the customer. Within Lean and Six Sigma, we call this Critical To Quality (CTQ, more on this later). Once it's clear what our customers want, we can continue with the Kano model to place and prioritize these customer wishes, so that we can get to work with the Voice of the Customer in a structured way.

The success of a product or service is ultimately determined by what the customer thinks and says. Three types of properties influence customer perception and are easy to trace with the Kano model. This model for customer satisfaction is a useful tool to develop precisely those product characteristics that make the difference. The primary or expected properties are so obvious that the customers do not even mention them when asked. Although they don't mention these properties, they indeed assume that they're met. If that's not the case, great dissatisfaction is the result. Performance characteristics are things usually expressed by customers when asked what they expect from a product. If the product performs better, its satisfaction will increase; if it performs less, satisfaction will decrease. Attractive features make the customer say "WOW!" They are often unconscious wishes or hidden needs. If they're missing, the customer does not miss them, but if they are present, customer satisfaction increases exponentially.

Chapter 7. Measure: Project WHY's, Data, and Defects

Measuring is the process by which numbers and symbols are assigned to concepts or objects in reality. This is done in such a way and according to agreed rules so that it's evident for all employees. Measuring is essential for gaining an objective understanding of what is going on, and how big, small, heavy, light, how much or how little a matter is. Measuring is central to Six Sigma because it's used as a tool to learn how causal relationships are. Measuring helps to learn what parameters influence the final result. People in the process must be allowed to perform these measurements. Please note that a distinction is made between measurements of the process and measurements that are focused on the *outcome of the process.*

To further illustrate this, take Andrew Johnson, a college student who wants to gain more weight or muscle. After a consultation with his dietician, he found out that he weighs 132 pounds, but needs to weigh around 176 to match his height. Together they make a program to increase his weight to 176 pounds. They make an appointment every other week for a consultation. At the start of the meeting, Andrew must always stand on a scale, and not only his weight is measured, but also his muscle mass and fat percentage.

These are the *result measurements* of his weight gain process; these results reflect the effect of the diet. But these measurements do not help with the weight gain process itself. After all, if you see no improvement (weight does not increase), you have no idea what's going on. That's why Andrew is instructed to keep a notebook in which he writes down how many calories he eats, how many exercise hours he does per day, and how high his heart rate is during these exercise hours. We call these *process measurements*. After all, this diet and life schedule is based on the knowledge that with proper exercise (weight training) and a more healthy calorie intake, Andrew will gain weight. This relationship between process and result can be checked in this way. At the same time, you will see that Andrew isn't going to structure his lifestyle based on weight, but according to his *process measurements*: calories and exercise. These are the parameters on which he has direct influence and which positively influence the ultimately desired result. Moreover, it gives the dietitian the possibility to check how Andrew's weight gaining process is progressing. The essence of measuring is that not only are results measured, but those process owners, in particular, do process measurements, and then specifically on those parameters that have a (proven) causal relationship with the result measurements.

What is the Measure Phase?

In the Measure phase, we determine what the measurement procedure is and how well the measurement system can measure the CTQs (Critical to Qualities). The team must ensure that the measurement system - and the data collected with it - is valid and reliable before continuing to analyze this data. To make a statement about this, a Measurement System Analysis (MSA) is performed on the measurement system. The Measure phase is often seen as the most challenging phase of a DMAIC project because reliable data is sometimes lacking, and it's challenging to implement an MSA. For Lean projects, it's critical to clearly define the definitions of the CTQ and the source for the data, so that there is no discussion about performance. The difference between the current performance (Baseline performance) and the desired performance is also determined within the Measure phase (Target performance).

Measuring is essential in Lean Six Sigma. Measuring can make or break your Lean Six Sigma project. If you don't collect any data, you will be confronted with disappointing results. Even if you win a little in the short term, the long run will catch on to you. The combination of data, knowledge, and experience is what ensures real improvement in a process. Measuring involves the following:

- You evaluate the existing measurement system and make improvements where necessary. If you do not yet have a measuring system, you need to develop or buy one.
- You collect data and observe the process.
- You map the process with more depth.

So, why is this necessary? You must be able to trust your data. It often happens that a team has spent a lot of time collecting data and then finding out that the measurement system used is unreliable. They discover, for example, that the clock was always switched on at different times in the process. As a result, they read a different turn-around time each time. Employees may misread an instrument or use different definitions of "defect." If you make your decisions depend on data, make sure that you can trust what the data tells you.

Furthermore, because DMAIC is a method based on data, you base decisions on facts and reality. People's opinions still count, but everything has to be balanced against what the data has to say to you. Also, you record what is really going on in a process so that employees know of each other what they're working on. Finally, you understand what needs to be improved and what doesn't. The key lies in what we have addressed earlier in this book, namely, out of the numerous activities your organization or department performs, only a handful of activities are of real importance to customers. It's your job to track down and improve these core activities. After that, you must eliminate as much work as possible without added value.

Data Types and Data Collection

In the Measure Phase, data is essential. Without data, the whole DMAIC model is useless. We distinguish two forms of data, namely qualitative or discrete data or quantitative or continuous (numerical) data. Both forms are used to collect data. Regarding qualitative data, proper measurements start by asking relevant questions, which means interviewing customers to work out what those questions are. By asking relevant questions, they can produce a rough prototype of a value model that has plenty of clarifying power; this, in turn, leads to the right decisions being made. The customers must be listened to when they talk about what they consider when they purchase a product or service from you. The primary focus should be on the customers in the relevant market segment. If you rely solely on your own internal judgments about the questions to ask, it will likely lead to weaker models and questionnaires that are just too long, and both of these are nothing more than a waste of time, money, and energy.

As far as qualitative research goes, this is about identifying what criteria are used by customers to evaluate their purchase options, looking at image, quality, and price. It is those criteria that the survey is based on, where customers are requested to rate how they felt your organization performed. This may be done over the phone, via the internet, or even in person, and that will depend on the size of the market and what products or services you offer.

There should be survey questions that verify the right person is answering the questions too. The survey scores are used in many different statistical tools for producing what is known as the Market Value Model. As such, the importance attached to image, quality, and price becomes fully visible during the value definition for the product and/or market; at the same time, the CTQ (Critical to Quality) factors are determined together with the importance of each. The result is a CTQ determination driven by data and facts that can, in turn, drive products, people, and improvements to the process through the entire organization.

Identifying Project Ys

Variation or spread is simply a deviation from the expectation. To further understand how to deal with data, I want to introduce a fundamental Six Sigma equation, namely: $Y = f(X) + \varepsilon$.

- First, we define "Y" as the result that you want to achieve.
- Second, we define "X" as the input. By the input or inputs, it's possible to get to the result.
- Third, we define "f" as the function. This is the manner with which your input transforms into the outcome.
- Fourth, we define "ε" (Greek letter epsilon) as the presence of defects or errors.

In simpler terms, the input is transformed by a process to the desired output.

To further illustrate this, take Amy, who wants to make cupcakes. She takes softened butter, eggs, self-rising flour, and the remaining ingredients necessary. Afterward, these ingredients are transformed by mixing and baking, to the desired outcome. The ingredients are the inputs (Xs), mixing, and baking is the function of the transformation process, and the resulting delicious cupcakes are the Y.

Not very hard math, right? Well, there's one more aspect, because, in practice, an outcome *always has a degree of uncertainty or variation*. It doesn't matter if the best chefs make the cupcakes or Amy makes the cupcakes; there's still some indefiniteness about the extent to which their actions produce the ideal outcome.

Just think about it. What if Amy didn't use enough softened butter or the oven wasn't set to the right temperature? Suppose Amy bake twenty cupcakes; would they all be ultimately the same? Nope, the cupcakes will- at least - differ slightly from each other. In Six Sigma, the small error that creeps in and causes this variation is represented by the "ε." These errors or defects can come from Amy herself (such as if she used the wrong temperature to bake the cupcakes), or be random (such as when the power goes out.) No matter the situation, there's always room for variation. So, let's take a closer look at it.

Variations and Defects

Variation or spread is simply a deviation from the expectation. To better understand variation, take Pablo, who flips a coin. How big is the chance that Pablo throws either heads or tails? If the coin is not spun, the likelihood is fifty percent. So when he tosses a coin twenty times, you expect ten times heads and ten times tails. Based on the probability, this is what we would expect. But is this true, or are we just assuming? Well, you can try it yourself; toss a coin ten times and see if you get five times heads and five times tails. Most likely, you won't. Why? Because there is *variation*, i.e., the output (the number of times head and coin) varies per series of ten flips. The degree to which your experience deviates from the expectation is the degree of variation.

When you accurately measure an output Y, you will see that it always varies no matter how hard we strive to make duplicates. Every output varies. Every single service or product that a business makes has differences in things like size, color, material, and degree of customer satisfaction. When you measure the event of something often, it will vary around the *average* value. Toss a coin often enough, and you see that the average of heads and tails tends to be around the 50/50 mark. When you measure the value of a given event, it will vary relative to the average. A top NBA player may have an average point per game of 29 over a whole season. However, in the game last Wednesday, he scored 25 points; this Wednesday, he scored 33 points - more than his average of 29. Why? These are examples of variation: the variation of the event relative to the average. The scope, trends, nature, causes, consequences, and control of this variation are the eternal obsession of Six Sigma. Nothing more is studied or treated than this in Six Sigma.

Measuring is the collection of data related to the input (the Xs) and the given outcome (the Y) that results from the process function "f". Measurements give you a quantitative understanding of the characteristics of the input and how this is related to the desired outcome. Measuring the input gives you a profile of the way a process runs concerning a goal or objective. Measuring starts with the Ys and then extends to the Xs to understand the causes. Let's make this more clear with an example. Mary, a college student, would love to have an extra $500 in her wallet, and this is her desired outcome Y. To measure her progress, Mary could check her purse every single day and count the number of dollar bills she has and finds she has too little compared to the ideal amount she wants. Mary could analyze the situation and conclude that the amount of money in her wallet is a function of how much she earns at her side job, how much taxes she pays, and how much she spends. These are the inputs (Xs). To have any form of influence, she needs to do something about the Xs mentioned to change Y. To bring about a change in the output Y, the person should measure and control the performance of the causal Xs. For instance, Mary could decide to find another side job to earn more or she could spend less money on buying things.

Unfortunately, most people never get past the Y. They look at it, like Mary does at the money in her wallet, and hope that measuring alone will bring about the needed change. Consider GardenJoy working harder and increasing productivity to improving results (the Y) without quantitatively examining which factors contribute to the success (the Xs of materials, supplies, level of quality, et cetera). Striving toward a goal without correct data - and in a disorganized and uncontrolled manner - most likely ends up in failure.

You can trace variation through data collection and statistics. If you then have insight into the variation, you can also identify where you need to improve the process to have as little variation as possible. This then minimizes the number of defects in your process, and it increases profit, the quality of products, and employee satisfaction.

Not all variation in the process falls within the upper and lower tolerance limits, and, therefore, there are defects. This means that there are variations in the process that are not acceptable to the customer and should not be present in the process. Let's say Julie buys a cellphone case for a 5.5-inch cellphone. Any variation with which your phone still fits in the case, for instance, 5.49 inch, will not be noticeable for the customer. However, we have a problem or defect if the phone no longer fits in the case; for example, if the case is 5-inch, the customer will not be satisfied. You can gain insight into this variation with the Six Sigma method. Based on these analyses, you can see how much variation is present, where this occurs most in the process and what causes it. That way, you also know where to adjust the process to minimize this variation.

But how do you know what is or is not acceptable to your customer? For this, you look at the Voice of the Customer (VOC): everything that falls within the limits of the Lower Specification Limit (LSL) and Upper Specification Limit (USL). On the other hand, you have the "Voice of the Process": the entire distribution (so also the part that falls outside the LSL and USL). The latter often expresses how the process works, but the VOC is most important because the customer comes first. With Six Sigma, we strive to have the entire process fall within the LSL and USL and, therefore, within the Voice of the Customer. The variation that is still present always falls within the quality limits of your customer. You've reached your goal when the variation present is entirely acceptable to your customer.

Chapter 8. Analyze: Finding Possible and Root Causes

After the Define and Measure Phases, we've now reached the Analyze Phase, great! But before we dive into the Analyze Phase, we need to take a step back, because we're at a critical point of the process. We need to look back at what we've done, so we get rid of wrong data, bad assumptions, and incorrect goals. These errors will snowball in this phase, and we will analyze with the false data and spending a lot of time, effort, and money, making unnecessary improvements. Before analyzing the data, we check the previous phases. What is our problem, again, as we looked at in the first phase? Are we addressing the Voice of the Customer? Know that the more data you have, the more resources you'll need in the Analyze Phase. So, be sure that you're collecting data on the correct variables so that the right variables are analyzed if you're going to tackle a process project. For instance, when you're looking at something like your variances in your processes - as far as the quality that's being produced, for instance, defect rates. This is more of a *quantitative study* and needs professionals with high statistics and Six Sigma knowledge. Qualitative studies, such as any project regarding waste, need people with a Lean skillset. In the Analyze Phase, the project team will focus on analyzing the sources of variation that were found in the selected process. Based on the problem, the techniques to find more root causes will be selected. Afterward, the project team will analyze the value stream, i.e., the set of activities that create value for your customers. Finally, the project members will also focus on identifying process drivers. These are activities that have a significant influence on the outcomes of processes.

What is the Analyze Phase?

In the Measure phase, we obtained more information about the current status of the process and the nature and extent of the problem. In the Analyze phase, you assess the ongoing process and try to find the leading causes of the issues identified based on further analysis. In this phase, you'll analyze all the information and data that you've collected in the previous stage. You will need this information to be able to find out the cause of the defects, lousy quality, and waste. To reach reasonable conclusions, a significant challenge here is that your team sticks to this data and does not rely on its knowledge and experience. Based on the data, you can conclude what the root cause of the problem is. The principal activities we undertake in this phase are:

- Searching for specific patterns in the data
- Identifying where a lot of resources, such as time, energy, and money, are wasted

How this helps us:

- They prove we need to identify the root cause, identifying the root cause *instead of symptoms*
- You'll find ways to speed up the process without compromising quality

- Also, you identify critical variables in the process you need to control, as well as essential factors in the process that must be under control

All possible causes are first collected based on the process and statistical data. You first identify all possible influencing factors. By further analysis and application of statistical methods, you try to limit the extensive list of possible causes to a small number of root causes (also known as the Vital Few or Red Xs). Two types of analysis are needed to discover and validate the root causes. In the first place, you take a closer look at the data that was collected in the Measure Phase. This analysis is called data analysis. You also investigate the value flow for possible waste (process analysis). Performing data analysis is also called the "data door" and analyzing the value flow the "process door."

With the data analysis, you verify the possible root causes with the help of statistical techniques, think of the following:

- Applying tables, graphs, and key figures to summarize and present the data.
- Regression analysis to investigate the relationship of influence factors on the result.
- Performing hypothesis tests to show which root causes are statistically significant.

In process analysis, the emphasis is placed on investigating the *efficiency of the process.* You do this with instruments that can be used to analyze risks, identify wastage, and make a distinction between activities that add value and actions that don't. It's often more Lean-oriented techniques such as Ishikawa, FMEA, Value Stream Mapping (VSM), Process Mapping, and the Five Whys method. The most important analytical tool for process analysis is the VSM. With a VSM, you visualize the process on paper and then assess for process steps, whether they are value-adding or non-value-adding.

Data and process analysis are aimed at understanding the problem/process and finding the causes of process variability. The aim is to determine what the leading causes are and how they affect the problem. Here you look for the causal relationships between influencing factors (Xs) and the output (Y), and you substantiate this with facts. The associations found between input (X) and output (Y) not only explain why the current performance is as it is but also form the basis for searching for the solution (s) in the Improve phase. In other words: the root causes found to indicate the short, medium, and long-term opportunities for improving the process.

Value Stream Mapping: Identifying Waste Causes

Value Stream Mapping is a commonly used technique to gain insight into wastes within a process. Business processes are mapped with Value Stream Mapping (VSM). The order of the activities that bring about a product or service is analyzed.

A value is added to each activity. VSM aims to improve the business process continuously. The business process is shown schematically in a diagram. The process steps and the information and material flows are visually represented (present state). From the diagram, it becomes clear where there are possibilities for improvements. With this tool, employees gain a shared understanding of the current situation and more understanding of each other's work and problems. Also, Value Stream Mapping is a useful model for working towards an ideal situation (future state).

Before mapping out the current situation, the purpose of the analysis must first be determined, for example, reducing the lead time. The goals must be acceptable to employees. The customer value is then identified per product. Then a team is put together, involved in the activities in the value and information flow. Take these points of advice:

- First, make sure you identify the actions (process steps) and assign the correct order. Frameworks determine in which role the activity is performed.
- Second, ensure that the processing, waiting, and recovery time of activity is determined. The extent to which an activity is carried out correctly in one go is expressed as a percentage.
- Third, the efficiency of the process is calculated (processing time versus lead time).
- Fourth, we now have a better view of the process whereby we can identify wastes much better.

The purpose of using this tool is to map the current situation and to identify wastes. As we explained earlier in Lean, there are seven well-known wastes. Some experts also add an eighth waste: Talent. The other wastes are: Waiting, Over-processing, Inventory, Transport, Defects, and Motion.

After this, you can work according to a step-by-step plan:

- Determine in advance of which product, product family, or service the Value Stream is captured.
- Describe the current situation (Current State) in a step-by-step plan, including every step (including delays) required to deliver the product or service.
- Describe the desired step plan in the desired situation (Future State).
- Analyze the differences between the two schemes and determine the action points based on this and set priorities to make the required improvements.

The Five Whys Method

The *five whys method* is a simple but very effective Lean Six Sigma tool to identify the root cause of a problem. The technique was developed by Toyota to perform root cause analysis for production-related issues. It's one of the tools available for implementing a root cause analysis. Although it's simple in practice, the output can be advantageous. Applying the method is done by these steps:

- Identify the problem for which the root cause must be identified. Ask the question, *why did this problem arise?* The result is an answer with a new or different problem.
- Take the first answer (or problem) and ask the question again: *Why did this problem arise?*
- Repeat these steps at least five times. If you think you can continue, by all means, continue until you reach the root cause of the problem.

Once during a Lean Six Sigma training, an attendee came up with a solution to the problem from the get-go. The problem was: "The customer receives an order late from the supplier for the fifth time." The attendee said straight away: "Then they must follow a time management course or hire more people." When dealing with organizational problems, never think of solutions straight away, because you've no clue if this is even the root cause. Thus, we start applying the *five whys method*.

- The first question would be: "Why did the customer receive the order too late from the supplier?" Frequently, the answer can be a real surprise, such as in this case.
    - The answer: "BlueTransport Inc., responsible for the delivery, did not have the correct address details of the customer."
- The second question would be: "Why doesn't BlueTransport Inc. have the correct customer address details?"
    - The answer: "The address on the shipment does not match the address of the customer?"
- The third question would be: "Why doesn't the shipping address match the customer's address?"

- o The answer: "The customer moved to a new city two months ago, and the new address is not yet included in the supplier's customer base."
- The fourth question would be: "Why is the customer's new address not yet included in the supplier's customer base?"
  - o The answer: "The only CRM system administrator was ill for the past two weeks, and no one has thought of changing the address."
- The fifth question would be: "Why didn't anyone think about changing the address?"
  - o The answer and root cause: "No one knew how to make a change to the customer base using the CRM system."

In the end, the root cause had to do with the fact that no employee knew how to change an address using the CRM system. This is the reason why the company failed to deliver the order. Without continuously asking "why," we may never know that this was the pressing issue. Now that we know the "real" problem, i.e., the root cause, we can take the necessary steps to tackle it and fix it.

Hypothesis Testing

We use tests to prove relationships between one or more Xs and the Y. To test and show (in a meaningful way) whether the suspected link between X and Y exists, it's imperative to know what you're testing. In the Lean Six Sigma field, hypothesis testing is used to make a statement about the existence of a statistically significant relationship between cause (x) and effect (Y) or between two or more causes. But what is a hypothesis? In science, a hypothesis is a proposition that has not (yet) been proven and serves as the starting point of a theory, a statement, or a derivation. Therefore, we're talking about testing a hypothesis because we assume that there is one. A possible relationship exists between two or more variables or that one variable (x) influences the behavior (the outcome) of another variable (Y). However, the statement is not (yet) proven, because it needs testing. To be able to test a hypothesis adequately, we will have to define a proposition. A statement that indicates that there's a significant relationship or significant influence between variables. However, if we make a statement that such a connection or difference exists, then the inverse statement is also true, namely that this relationship or difference doesn't exist. Therefore, we always define two propositions with hypothesis testing, namely the so-called null hypothesis and the alternative.

- The Null Hypothesis (H0): there's no significant difference or relationship
- The Alternative hypothesis (Ha): there is a meaningful relationship or difference

Furthermore, there are a few critical elements for formulating a reasonable hypothesis, namely:

- **The team is suspecting a relationship between the variables.** Hypothesis testing is, of course, not arbitrarily, but based on suspicion of the existence of a particular relationship between variables.
- **They are statements, not questions.** It is crucial that a reasonable hypothesis is defined as a statement *and not as a question*. By drawing up a null hypothesis and an alternative hypothesis, we describe both possible outcomes, and it's up to the team to show which of the two is appropriate.
- **It can be tested/measured.** And last but not least, it is, of course, important that the hypothesis can be examined, which means that we must have measurement data by which we can say whether there is a relationship or difference or the absence thereof. Hypothesis testing makes sense when we are not examining the entire population, but have taken a sample from the people in our audience.

There are a few steps you can take to test hypotheses, whereby we go from a practical problem to a statistical problem, to statistical solutions, to practical solutions.

- First, define the research question. The research question is a direct consequence of the issue from the project charter. The Project Charter is a contract between client and contractor: a communication document that the sponsor and project manager agree upon at the outset.
- Second, define the null hypothesis H0.
- Third, define the alternative hypothesis Ha.
- Fourth, determine the significance level.
- Fifth, collect measurement data and rate and test the data.
- Sixth, run a test with a statistical program like Minitab or SPSS and draw the statistical conclusion.
- Seventh, write down the conclusion in a practical way, so that all stakeholders can understand it.

Take, for instance, OnTheWay, an organization for traffic and tourism that's researching the ease of parking of two types of cars, namely electric cars and gasoline cars. This research is conducted - under the same circumstances - by 30 test subjects by having each person park both types of cars once. They defined the following research question: "Is the average parking time for both types of cars the same?" Afterward, they established hypotheses. H0: "There is no difference in the time required to park the cars," and Ha: The average difference to park both cars does exist." Then, they determine the significance level. Because this is a standard test whereby defects don't lead to severe consequences, a reliability level of 95% is acceptable. Then, OnTheWay collects the number of seconds it takes for the subjects to park the cars. Afterward, the rate and test the data by using a Paired t-test in the statistical program SPSS by IBM. Last - but not least - they write an informational and concise report of their findings in plain English. With this report, they can continue to the next phase, namely the improvement phase.

Let's take a look at what this phase is all about!

Chapter 9. Improve: Generating Solutions

Now that we have a good overview of Lean, Six Sigma, and Lean Six Sigma, it's time to focus on the practical side of Lean Six Sigma. In this chapter, we take a closer look at the Lean Six Sigma process and how you can implement the same. As explained previously in this book, there are two critical methodologies for implementing Lean Six Sigma in your organization, namely DMAIC and DMADV. In this chapter, you'll learn more about both these standard processes and their benefits, so that you can decide the process you'd like to follow.

What is the Improve Phase?

In the Improve phase, you conduct changes to the process so that the found defects, wastage, and extra costs – related to the needs of the customer and as identified in the Define Phase – are eliminated. This link with customer needs is essential. After all, you only work on the (root) causes that touch on the problem or need. These are the causes you identified in the previous phase, Analyze. In the Improvement Phase, you focus on the following:

- You use tools and techniques to channel creative thinking and formulate adequate solutions to the related root causes. You no longer blindly follow what management demands, but do your thinking based on the gathered data.
- Besides, you strive to learn as many best practices as possible. Many DMAIC projects have been conducted in all kinds of organizations. So, why not learn from others before trying things yourself? This will fast track your Lean Six Sigma projects.
- Also, in this phase, it's emphasized that you develop criteria with which you can select solutions. You determine the course to be followed that belongs to the chosen resolution and plan the full implementation thereof.

Always keep in mind that the objective of the Improve Phase is to *implement and verify the solution* to the problem. To determine the optimal setting for a process, specific techniques can be used, such as regression analysis or cost-benefit analysis. By adequately going through this phase, you don't get caught up in figuring solutions that don't apply to the found root cause or don't suit the available resources. Also, you'll develop more creative solutions that are related to the root cause and are proven in the past. With experience and gaining more knowledge of real case studies, you can substantiate why one solution is advised above another solution.

Steps Toward Improvement

In this phase, we'll identify, prioritize, and implement the improvements for the most critical inputs and outputs we discovered in the previous phase. The corresponding statistical formula is: $f(X1 + X2...) = Y$. The output we strive to attain is, of course, the fact that we've identified possible solutions and implemented the best to eliminate or, if not possible otherwise, reduce root causes as much as possible. Usually, we make an improvement plan to ensure proper implementation. Key questions surrounding the plan are: "What are we improving?" and "In what part of the process will we embed the solution?" After the plan, we move on to implementing a pilot. We never embed an improvement organization-wide from the get-go. Instead, we figure out parts of a process where we can test the improvement. If the pilot is successful, slowly but surely, the improvement can be embedded into other parts of the process or other processes. Performing a pilot is crucial because it gives us crucial data to work with. If the pilot fails, we return to identifying improvements and test another one.

We can distill this phase in a couple of steps:

- First, make sure that the project team has reviewed all items that came about in the Analyze Phase. Check created documents, gathered data, and possible errors.
- Second, discuss with the team members an improved version of the business process you're tackling. This needs to be a process that would be wholly optimized for customer satisfaction.
- Third, it's advisable to make a transition plan. This will facilitate the move from the current process to the newly crafted process when doing so, take the PDCA (Plan, Do, Check, Act) cycle in mind.

Cost-Benefit Analysis

A cost-benefit analysis (CBA) is a systematic approach to estimate strong and weak points in, for instance, transactions, investments, business processes, or other activities. The predominantly monetary evaluation method is used to identify effective options and make responsible choices that both offer benefits and are cost-effective. However, in this process, values are also given to intangible issues such as the benefits associated with working in a specific area or possible loss of reputation after making risky strategic choices. The income and expenses are therefore expressed in terms of money and are adjusted for the time value of money. This is needed to use the net present value frequently. The cost-benefit analysis is used, among other things, for making purchase/production decisions, for example, in real estate investments, but also in decisions that affect society. In a social cost-benefit analysis, the external effects of a decision are also included in the evaluation. However, in most organizations, the focus is strictly on the cost benefits. There are various advantages this analysis brings to the table, such as the research into the viability of a project proposal, the evaluation of new purchases/investments, and the assessment of the desirability of a proposed policy.

These are the steps to make your cost-benefit analysis:

- **First of all, you need to determine the expected costs and benefits of the project.** Take the time to brainstorm about the costs associated with the project and compile a list of all possible costs. Do the same for all the benefits of the project result.
  o Some crucial questions that can be asked here are: Are there unexpected costs that may still be identified in advance? Are there advantages associated with the result that was not initially assumed?
  o Think of different forms of costs, like operating costs, personnel costs, real estate, facilities, and material costs. But also unnecessary costs such as time, energy, or a loss in customer satisfaction.
  o Types of benefits are higher revenue, more savings, better customer satisfaction, and more satisfied employees.
- **Second, express the costs and benefits in the same unit.** Where costs are relatively easy to communicate in terms of money, the benefits are slightly different. However, both costs and benefits must be expressed in the same unit since they must be compared. Therefore, think about the effects of the benefits that the project brings forth.
  o Related questions are: Is an increase in customer satisfaction to be expected? And will this provide

extra revenue? A monetary value must be assigned to all these benefits.

- **Third, compare the costs and benefits.** In the final step, the value of the costs must be compared with the value of the benefits. To do this, calculate the total costs and the total benefits and compare these two values to determine whether the benefits outweigh the costs. Then make a decision based on the results of the comparison.

To further illustrate this take, BetterTextile Inc., an Eastern European manufacturer of textile. BetterTextile Inc. notices that earnings rose for the third consecutive year and has insufficient capacity to meet the rising demand for their textile in Western Europe. BetterTextile Inc. is considering tapping into a new market in Western Europe and requires more employees for this. The business premises must also be expanded, and the new employees will be provided with advanced tools and better training to hone their skills. Thus, BetterTextile Inc. made a cost-benefit analysis to determine whether the new strategy is worth pursuing. In this example, we assumed the following:

- BetterTextile Inc. expects to earn back the investment that they made after one year.
- The company expects productivity to increase by around 6% with better-trained staff using more advanced equipment.
- Also, BetterTextile Inc. expects revenue to increase by 30% after expanding capacity.
- Finally, BetterTextile Inc. hires additional staff for 200 hours per month for 40 dollars per hour.

When we take the benefits of $300,000 and costs of $210,500, we can say that BetterTextile Inc. will make a profit on this move. Of course, in real life, these assumptions have to be validated in more detail and with real data. But this will do for an illustration. The illustration shows that making such an analysis helps in making moves that may seem bold, but can turn out to be very profitable.

Solution Parameters and Generating Possible Solutions

A parameter is a letter in a formula that represents a constant. It's like the creator of the assignment doesn't want to reveal this continuous. Therefore, a parameter is something different than a variable. A variable is x or y: a letter that assumes all kinds of possible values, and for each of those values, you get a point on the graph. You can plot a variable on the x-axis or the y-axis. A parameter has only one constant value, *but we don't know it yet*. What are the consequences of the graph? If you have a function regulation with a parameter in it, then you have a problem because you can't draw points (x, y), you cannot enter a formula in your graphing calculator. The only thing you can do is replace the parameter with a number you want, like parameter = 1. Then you have a formula without a parameter, and you can draw it. Then choose another number, parameter = 2, and draw the graph again. You can continue for a while, and then you get a collection of a lot of graphs. This is called a chart bundle. A function description with a parameter, therefore, never includes one graph, but always a graphing bundle.

In Lean Six Sigma, we see something similar. In the Improve Phase, we tend to make use of so-called solution parameters. These solution parameters help us select the right improvements to solve an issue. The first step in doing so is by developing a *decision statement*. Utilizing brainstorming, we can gather various solutions to clarify the purpose of the decision the team is going to make. While forming your decision statement, think of matters like how you'll manage customer expectations and how you'll react and evaluate to possible decisions the team took in the past. For the second step, you need to formulate a list of around ten criteria to help you solve the issue at hand. With criteria, we can look at the list of possible improvements we've generated and tested them against the criteria. The criteria can be divided into two forms, namely the "Musts" and the "Wants." For the last step, refine the solution criteria as much as possible before you use them to select the right solutions or generate other solutions. New solutions could be made, for instance, by combining two solutions or improvements you found. All team members should be aware of the implications of the criteria; there should be no room for ambiguity.

It's not always easy to figure out innovative ideas; humans often need a little push to wake up our creative sides. The SCAMPER technique is a useful brainstorming technique to help us do just that. With this method, one can produce ideas for new products and services by asking various questions. The questions generate creative ideas for developing new products and improving current products.

- First, select an existing product or service.
- Second, asks questions by using the SCAMPER method. It's most useful to do this with the team.
- Third, categorize the answers into something like "useful," "maybe useful," and "not useful." Or, you can use something like the Likert scale to generate the best solutions.

SCAMPER is an acronym for Substitute, Combine, Adapt, Magnify/Modify, Purpose, Minify/Eliminate, Rearrange/Reverse. These are questions you can ask for each element:

- Substitute:
  - Are there materials or resources that we can substitute to improve the product?
  - Can we use the product/service for other purposes?
- Combine:
  - What will happen if the product/service is combined with another product?
  - Could the product/service be used for a different purpose?
- Adapt:
  - What other context would the product/service fit in?

- o How can the product/service be adjusted so that it runs better?
- Magnify/Modify:
    - o How can we adequately modify the shape or appearance of the product/service?
    - o What can be added to the product/service to make it better?
- Purpose:
    - o Are there other people than our current customers that will benefit from this product/service?
    - o Can the waste from this product/service be reused?
- Eliminate/Minify:
    - o What would this product/service look like if it was simplified?
    - o Which functions or components can be omitted?
- Rearrange/Reverse:
    - o What happens if parts of the product/service are assembled in a different order?
    - o What happens if the product is turned over?

Chapter 10. Control: Sustaining Improvement

After the Improvement Phase, it's finally time for the last phase of the DMAIC cycle, namely the Control Phase. The purpose of the Control phase is to help attain the desired results in the organization. Although the problem may already be resolved, the team must not forget to take this phase seriously and prevent the problem from reoccurring. The best way to achieve this is to guarantee improvement with "Poka Yoke" solutions, which are not dependent on the employee. This is not always possible, and then the solution still depends on the employee's working method. This is particularly the case if the solution results in a different method. Possibilities for guaranteeing an improvement, in this case, are the drafting of new work instructions, the provision of training, and the adaptation of quality documents such as the Control Plan. In the Control phase, the team also figures out how much resources were saved, and this is compared with the expected savings at the start of the project. Afterward, the order is formally returned to the client, the

Champion. After all, the job of the Champion or the department manager is to value the team for the performance delivered.

What is the Control Phase?

In the control phase, the goal is to anchor the implemented changes and ensure that the issues don't return. To achieve this, you monitor the most critical x's and y's with the help of a control chart. Another tool that is often used in the control phase is a Control Plan. These techniques and tools all help with the essential checking we need to do. Why do we even bother checking? Well, because we want to be sure that the progress we made is lasting. Your employees need procedures and tools to support changes in the way they perform their work. The team must pass on what they have learned to the process owner, and they must be sure that all parties involved in the process are properly trained to handle the new procedures. Therefore, a few key activities take place in this phase:

- First, we ensure that new or improved procedures are documented adequately.
- Second, we make sure that every related employee gets the necessary training to work with the new procedures.
- Third, we set up procedures to detect signals of things going wrong in the process, for instance.
- Fourth, the daily management of the new/improved process is delegated to the process owner, who will then oversee if the process works as planned.

If the team did a good job, this would result in many benefits. It will prevent relapse because changing people's habits means more than just turning on a switch. The above action points should make it easier for employees to get used to new procedures and not fall back in their old habits. Furthermore, it helps you respond swiftly to future problems. If you follow essential signals in the process carefully, you will be able to react quickly if new issues arise. The faster you respond, the higher the chance that you will discover the root cause and find a solution. Finally, it creates a culture whereby we are continuously learning and improving. There is a chance that other people in your organization will do the same or similar work you've done. If you document the work you've done, the obstacles you faced, and how you overcame them, others in similar roles can learn from this.

The Control Plan

Being in control is essential when changing or improving a business process. It's no coincidence that the final phase of the DMAIC cycle is the Control Phase. You want to go through this phase to guarantee that the new working method is followed so that you do not fall back (consciously or unconsciously) in the old way of doing things. In practice, frequently, we see this relapse happen as soon as there is pressure or stress. The old way of working is familiar, and the outcome is known. It's precisely in these situations that it's essential to monitor the process to prevent colleagues from falling into old behavior or bad habits. All operational elements that we want to watch and control come together within Lean Six Sigma in the Control Plan, also known as the assurance plan. For each component, we indicate which concrete indicators we want to monitor, how we are going to do this, and what we do if the indicator goes beyond the set limits.

So, how do we make a control plan? To make one, take a look at these points and the related example:

> • **Describe the parts (Y) of a process you would like to control.** Take InnoSub, a technology advisory firm that helps small businesses gain subsidies for working on innovative technologies that serve society. For this first step, the subsidy application process of the firm has been adjusted so that the time from "request" to "pending request" is within one week.
>
> • **Then, determine the measurable indicators (Xs) of this selected component.** For InnoSub, there are a couple of

quantifiable indicators. Think of processing time of checking subsidy documents, amount of available employees, number of cases in the queue. You work out steps 3-5 per indicator; for this illustration, we take the "number of cases in the queue."

- **Now, you can prepare your measurement plan that includes how you'll measure the performance of the process and who will do the measuring.** InnoSub's Subsidy manager measures the "number of cases in queue" twice a week, on Monday and Friday at 10 pm.
- **Set related boundaries: when should action be taken?** The Lean Six Sigma team who worked on improving this process concluded that action must be taken if there are more than 15 files in the queue and no employees available to take these on.
- **Determine what you do if the indicators exceed the limits employing an OCAP (Out of Control Plan).** When things go wrong, InnoSub figured out a counteraction: call in the help of a few remote contractors to help deal with the remaining files and bring it down to at least 15. It's better to get the number back to zero as much as possible, but it cannot exceed the 15.

Control Chart

The control chart is the heart of "Statistical Process Control": Keeping a process under statistical control. The control chart is a graph that shows trends, shifts, or patterns in the output of a process over time. The purpose of the control chart is to discover whether the process is stable and under control. Take GreenLeaves, a company based in London that produces tea bags. Let's say the weight is measured for numerous tea bags; this weight can then be plotted in the chart. Essential to the control chart is the average ("centerline") and the so-called control limits we discussed in a previous chapter, namely the upper control limit (UCL) and lower control limit (LCL).

These are the limits that indicate how much the process may deviate from the average. For this example, let's take a UCL of 2,5, an LCL of 1,4, and an average of 1,7 grams. The distance between the control limits indicates how much variation in the process can be expected. Observations that fall outside this space are "outliers," these are matters that demand further investigation. For instance, if GreenLeaves were to make tea bags of 3 grams, these would be outliers, because it's above the UCL of 2,5 grams.

Earlier, we addressed normal variation and the unexpected, unique variation. The latter is a variation that occasionally occurs and can break the process you're controlling. This is due to the influence of a demonstrable cause. For the tea bags that weigh too much, there is probably an unusual variation. Something in the process has ensured that these tea bags are too heavy. When we've figured out there are outliers in the process we're attempting to control, we can find the root causes of these with various Lean Six Sigma tools, like the Five Whys method we already addressed or the Ishikawa diagram. Besides, it's fundamental to look for patterns in the data. These patterns can help you be in "real control" of situations like these popping up. The ideal situation is that you intervene at the right moment and can foresee an outlier from rising to the surface.

The control chart is an example of a tool that can be used in more than one phase. It can be used in both the Analyze Phase and Control Phase. In the former, it's used to investigate whether the process to be improved is stable, and in the latter, it's used to recognize when a process shows signs of losing control. It's a great tool because it says a ton about the process you're controlling. However, the main disadvantage is that it doesn't say anything about whether the process meets the needs and/or desires of your customers. Why? Because the control limits (UCL and LCL) are often attained by using statistics and aren't the measures used by the customer.

Now imagine that the customer of Greanleaves wants smaller tea bags, with a minimum weight of 0,7 grams and a maximum weight of 1,2 grams. Our process will not be able to meet the wishes of the customer since our tea bags vary between 2,5 and 1,4 grams. As I mentioned multiple times throughout this book, Lean Six Sigma projects are all about striving for processes that meet the needs and desires of customers. Without a control chart, you're left in the dark regarding the organization's current process performance. By figuring out a way to deal with the inevitable variation and then improving processes by listening to the customer, we create a robust process that puts a smile on the customer's face.

Mistake Proofing: Poka Yoke

Poka Yoke is a Japanese term that means "error prevention" and was developed by an engineer at Toyota in the 1960s. Poka Yoke is used to prevent and resolve defects in the production process so that quality control afterward is no longer necessary (or to a far less degree). In Lean Manufacturing and Six Sigma, in particular, Poka Yoke is one of the most common methods for making sure that production runs smoothly from start to finish. What does Poka Yoke mean? "Poke" means "unintended error," while "yoke" means "to avoid" or "occurrence" in Japanese. By using Poka Yoke, mistakes are pretty much impossible to make, and the result is the right action being forced, ensuring there are no misunderstandings. It's all about using measures that stop mistakes from happening.

Poka Yoke includes many simple solutions that are also useful and cheap, and they are easily integrated into your design or one of the intermediate steps. Perhaps one of the best examples is a mobile phone SIM card. It can only go into the phone in one way, and that means there is no room for error. Poka Yoke is one of the most powerful tools, warning you of errors and allowing for quick reactions to stop deviations. It fits perfectly with Lean Six Sigma and, as such, is worth a look in terms of the DMAIC phases:

- **Define** – this phase involves describing and defining the problem that is causing the error or defect. The description is objective - with no direct conclusions. In the production process, the workplace can be observed, also called "Gemba," the Japanese word meaning "actual place." In your context, that could mean the factory, the manufacturing plant, and so on. The process runs in the workplace, and there is every chance that what is causing the problem is hidden. If the user is causing the problem, that problem should be defined objectively from the user's point of view.

- **Measure** – This phase tends to be applied to the more complex issues in production. Tests determine how much this problem happens, and that is then converted to a percentage. The higher it is, the more important it becomes to tackle the problem and solve it where it is caused. As well as production errors, user errors can also happen. In this case, test-groups are deployed for testing the product for a set length of time. The outcome is used to solve the problem.

- **Analysis** – this phase shows if you can apply a Poka Yoke measure. The process goes through a thorough analysis to trace where the defect is, and it is only then that a solution can be devised.

- **Improve** – the analysis provides the information needed to tackle the problem, devise the solution, and implement it. Many times, a Poka Yoke solution may be applied to great effect, ensuring that the mistake will not be made again.

- **Control** – in the final phase, the adjustments are looked at, and the effects measured. If you use a Poka Yoke measure and it works well, eliminating the risks of other potential errors, you get the "Zero Quality Control," meaning the requirement or personal inspection is removed because the potential of human error is removed.

There are two different Poka Yoke solutions for solving production problems:

- **Visual Aids/Steering Mean** – these are visible, showing the method clearly. Think of a warning traffic sign or a pictogram. The steering revolves around behavior and deviation warnings. Going back to that warning traffic sign, it may light red when a driver is going too fast or go green when the driver is going at the correct speed.

- **Compulsory Means/Fail-Safe (SF)** – with this one, users are forced into doing or not doing something. Let's say a highway has been closed for maintenance; message signs are used to force drivers down from three to two lanes of traffic, and red crosses are used to indicate the closed lane.

To further illustrate this, take the following example from HelloWeather, a company focused on making eco-friendly rain gear. Customers have complained that after a couple of weeks of use, cracks appear at the bottom of their newest Hello Raincoat. HelloWeather follows these steps to fix the problem the Poka Yoke way:

- **Define.** We start by objectively defining the problem. When putting on and taking off the Hello Raincoat a couple of times, cracks appear. The definition can be used to see how the suit is put on and taken off and in which order; first, insert the right arm and then the left arm (or vice versa), pull the zipper from bottom to top, pull the hood over the head, etc. Taking the coat off is in the reverse order.

- **Measure.** A test group can put on and take off the Hello Raincoat a few times a day, for several days. It can then be shown in samples how many of the test subjects had a coat that had cracks after the test period.

- **Analysis.** To know where the error is, it is wise to see what exactly happens when the coat is put on and off by the test group. The problem seems to be with the zipper. The light polyamide material that the coat is made of is not flexible enough when the coat is put on or taken off. As a result, there is pressure on the zipper, especially on the lower part, which quickly creates cracks/holes at this location.

- **Improve.** Now that we know how the holes appear at the bottom of the zipper, we can look at adjustments. The zipper is traditionally placed vertically in the suit. The solution is more straightforward than expected. By sewing the zipper diagonally in the suit, there is less pressure on the underside of the zipper when getting in and out. This reduces the chance of cracks/holes.

- **Check.** The suit can be taken directly into production and offered for sale. To be fully assured that this is the solution for the cracks/holes at the bottom of the zipper, a test group can again be used, exactly as we did in the Measure Phase. With Poka Yoke, it's not directly about the test and measurement results, but the *real solution that is found.*

Chapter 11. Lean Six Sigma with Agile and Scrum

Since the time of Henry Ford, manufacturers were keen to realize more effective and efficient processes. With the rise of various car manufacturers, such as Toyota, new methods were developed to squeeze as much productivity in processes as possible. In the past, organizations assumed they worked efficiently and effectively enough. The opposite was the truth. They used to work with something we now see as obscure and highly ineffective and inefficient, namely the waterfall method. To tackle projects, they focused on various phases, namely first gathering requirements, performing analysis, designing the product, creating/coding it, testing it, and operating it. The major flaw in this system is that each phase in this method is completed in a vacuum! This means that professionals of all stages don't communicate with each other, and an organization only moves forth when a phase is completed (which is never the case in the dynamic world we live in!).

There was room for lots of improvement, and luckily this improvement came about when Hirotaka Takeuchi wrote a paper wherein he addressed the crux of what we would now call an *agile methodology*. It didn't take long before more agile methodologies popped up in different branches, such as Extreme Programming (XP), Kanban, and Scrum. It was only in the year 2001 within software development that a group of professionals came together to write the "Agile Manifesto." They included the basic principles of Agile working extracted from the various agile methodologies that were already practiced. Agile is now spreading rapidly to all types and parts of organizations. There are already hundreds of thousands of Agile practitioners around the world, and this number is only increasing. Especially Scrum has made a tremendous rise in organizations of different sizes, in various industries, in different countries all around the world. Let's take a closer look.

Agile breaks up significant product developments, in short, well-arranged periods (iterations) of two to a maximum of four weeks. Those iterations are small stand-alone projects that are managed by Timeboxing. The Agile approach enables a project team to quickly adapt the project to a changed situation or desires of the customer. Agile is the approach that is suitable for the dynamic world we live in. This is in contrast to a traditional project approach, in which a team tries to avoid change as much as possible by, on the one hand, laying down the specifications in detail and, on the other hand, setting up a formal process. People who flourish in an Agile team are people who are not afraid of change and are less inclined to look for certainties. Being able to start something without having a well-known result is very important. Agile is not only suitable for relatively simple business projects. Agile can be a perfect companion for complex and long-term projects too. Agile splits complex projects with iterations or sprints, making them more manageable. If objectives or circumstances change during the project, this is anything but a problem within Agile. It's essential here that the organization in question is one with an open, communicative culture. Scrum is easy to apply, and there is hardly any overhead. A Scrum team is self-managing and involves every team member in a project, including the customer, the user, and the client. Fast, clear insights, and clear expectations for all parties, deliver the desired product at the right time.

Agile and Lean Six Sigma are two philosophies with similar ideas. Agile is short and cyclical, which means that it is possible to anticipate market demand and customer wishes faster and better. Also, Agile contains the tools to repair a defect if it occurs quickly. This is because a 2-4 week cycle (the sprints/iterations) is used instead of a monthly or annual cycle. The idea of a Lean Startup outlined by author and entrepreneur Eric Ries also fits well with Agile thinking. In the book The Lean Startup, for example, the Minimum Viable Product (MVP) is mentioned. This (part of a) product already works and meets specific minimum requirements, but is not yet perfect or complete. This method can enormously shorten the "Time to Market." This is because the customer immediately receives certain functionality to get started. An additional advantage: feedback from users returns faster and contributes to the further development of the product.

Just like Lean Six Sigma, Agile pays a lot of attention to (customer) value and errors. A critical Lean Six Sigma principle is: "never pass a defect!" For example, FMEA (Failure Mode and Effect Analysis) is a technique that assesses the risks and effects of failure within the Lean philosophy and formulates countermeasures. Also, "Poka Yoke" - think of the SIM card that can only enter your phone in one way - is an important principle to prevent errors from the design itself. In the beginning, Agile projects focus a lot on business value. In Lean Six Sigma terms, this is called the "Voice of the Business (VOB)." Besides, applying FMEA makes it immediately clear what the customer finds essential, within Lean Six Sigma, the "Voice of the Customer (VoC)," but also whether the objectives are achievable within the set time. Besides, Agile is also focused on visualizing the progress of the "User Stories," such as requirements from customers. All ties could even tie in with another Agile methodology, such as

Kanban. The Kanban methodology makes visualization possible, a technique that is also frequently used within Lean Six Sigma.

It sometimes seems that Agile, Lean Six Sigma, and Scrum are entirely different methods and that you should opt for a specific technique. The opposite is exact: the techniques overlap and complement each other. The trick is to get the items that work best for your organization from the different approaches. But how do you know which parts you could best use in your organization? It is, therefore, necessary, *before you start,* to exchange ideas with a consultant or professional in your organization.

Lean Six Sigma is usually used in standardized environments, such as production processes. Scrum is widely used for creative and innovative projects, no matter what industry. But with this separate approach, you exclude projects and processes that can benefit from a combined approach. In a production environment, for example, you can let the development and production of a product go hand in hand by combining Lean Six Sigma and Scrum, for example, in the form of Extreme Manufacturing. A good example is the Toyota Prius. Toyota, the birthplace of Lean, has used Scrum to develop its innovative hybrid car. The Prius was delivered within 15 months, from idea to working car. How have Lean Six Sigma and Scrum strengthened each other here? Lean elements came back in placing customer demand first and learning from mistakes. In Lean environments, the discovery of a fault means that you halt the production process and immediately resolve it as a team. At the same time, the Scrum approach helped in dealing with the uncertainties in the development process, which they were able to navigate thanks to the sprints.

Scrum provides practical tips for Lean Six Sigma principles. Scrum creates, among other things, flow and pull through multidisciplinary teams, quick deliveries, and the incorporation of interim feedback. Thanks to the sprints, you can continuously check whether you are still delivering value for the customer. Take this practical illustration of both methodologies:

- Lean Six Sigma principle #1: Specify a value for the customer.
    - Scrum approach: This principle ties in nicely with the fact that the Product Owner (one of the most critical roles in Scrum), has the important task of continuously translate the customer wishes to a backlog with related tasks to fulfill these wishes. These tasks are then prioritized, tackling the most valuable items first.
- Lean Six Sigma principle #2: Identify the value stream.
    - Scrum approach: The complete value stream is combined in one Scrum project team. As a result, the waiting time is drastically shortened, and it's possible to respond to pending work with more flexibility.
- Lean Six Sigma principle #3: Create more flow in processes.
    - Scrum approach: Through intensive cooperation in small teams that work with short iterations, fast delivery, and feedback moments, waste such as transfer, waiting, errors, inventory, and overproduction are prevented.
- Lean Six Sigma principle #4: Create more pull.
    - Scrum approach: Working in sprints of approximately 1-3 weeks ensures rapid delivery and flexibility. You can respond directly to dynamic customer demand.
- Lean Six Sigma principle #5: Strive for perfection.
    - You get perfection through continuous learning and improvement. After every sprint, there is a review in which the product or service and work process are evaluated. The purpose of this is to make the process run even better in the next sprint.

All in all, it's precisely the combination of Lean Six Sigma and Agile methodologies, such as Scrum, that provides dynamicity and flexibility for your organization. I would, therefore, advise you to reinforce one approach with another whenever you can.

Chapter 12. Mistakes to Avoid in Lean Six Sigma

Anything worthwhile takes time. And anything worthwhile takes failures. Although failure is inevitable to run adequate Lean Six Sigma projects, it can also be very beneficial. It can reveal where there's room for improvement, give us ideas for what to try next, exposes our weaknesses, and grants us opportunities we wouldn't have thought about. No doubt, having clear goals and entering Lean Six Sigma projects with a positive mindset is vital to achieving success. Although failures may be inevitable, it doesn't stop us from learning from them and learning from other people's faults. This can help us not to make the same mistakes again or not make the in the first place.

First, avoid analysis paralysis. During the analysis phase, you may tend to keep going and going with, for instance, the root cause analysis so that you lose sight of the most crucial reason for the improvement project, namely, the critical reason for making a difference and seeing the positive opportunities for your company. It's vital to limit the scope of the project and to prevent you from taking all side roads available to you. Save these options for future projects, but keep the original scope. I know it can be challenging to determine when you should stop analyzing and start improving. But try to see this decision as an impartial decision and weigh the pros and cons. You are ready to move to the next phase if you can ensure that you know enough about the process, the problem, and founded root causes to come up with useful, innovative solutions. The project champion, the project supervisor, has an important role. He/she must keep an eye on company interests when answering these questions and keep the team on the right track.

Second, don't get overwhelmed. Achieving Six Sigma (3.4 defects per 1 million options) can be a beautiful end goal, but with one project, you're not likely to achieve anything near this precision. In Lean Six Sigma projects, the usual trend we see is to go from two sigma to three sigma and then to four sigma. Small, bite-sized projects ensure that your performance is moving in the right direction, so be prepared to accept a small increase in the sigma value of the processes, but always remember that these modest increases, day in, day out, will help you win in the long term.

Third, don't *just* conclude! Many managers tend just to draw conclusions when they face a problem. Sudden decisions can be costly and be an obstacle in addressing the cause of the problem. Doing something randomly, or just calling something randomly in business, without collecting and analyzing facts and data is not the best approach to solve complex business problems. Lean Six Sigma involves understanding what the problem is and then going through some steps to understand it better (define, measure); dig deeper to found the root cause (analyze); look at the different solutions and then choose the best (improve) and implement this solution and retain its benefits (control). Although this approach sounds very simple and meaningful, it feels unnatural to many executives who think they know everything best. Look at the enormous amount of decisions that are taken every year about production systems, warehouses, office spaces, IT systems, company reorganizations, new products, organization training programs, and more. These decisions are often random "solutions" that popped up during an executive meeting. But companies often find out half a year or a year later that these "solutions" do not meet the needs of the customer.

Fourth, thinking, "we know it all already." A brief look at books about Lean Six Sigma or a quick look at your processes can make you feel that your organization "does it already." Many managers think that their problems will solve themselves if they use a systematic problem-solving process. But often they don't think about solution options before they implement them. Or they do not sufficiently test the different options. You may think, "We are already using flowcharts or root cause analysis methods." Many organizations use these techniques, but often without first understanding the real requirements of the process from the customer's perspective. If you do not implement Lean Six Sigma in a structured way, you will never fully use the power of process analysis techniques to find out how existing processes work. Real support from management for such a process also analyzes rare but very much needed. This creates a bottleneck in many Lean Six Sigma projects: the employees want to change with Lean Six Sigma, but the management wants to do "it" how it was always doing "it," whatever that's supposed to mean. A well-designed Lean Six Sigma program is based on existing knowledge, places improvements in a framework in which everyone is involved, and introduces a comprehensive set of tools for the entire organization.

Fifth

, embracing false beliefs about Lean Six Sigma. If you want Lean Six Sigma to work in practice, you will have to eradicate some false notions, namely:

> - **Thinking that Lean Six Sigma is only helpful when an organization wants to improve production processes.** The methodology is not necessarily exclusively related to production processes, and this is proven by various

companies that deliver other services than manufactured products, such as Software as a Service.

• **Knowing that Lean Six Sigma is everything we need.** As we mentioned above, doing things in a vacuum is never the right decision. The same goes for Lean Six Sigma, and this approach works best with other methodologies, such as Scrum. Take the best from these worlds to get the projects off the ground.

• **Thinking that Lean Six Sigma is nothing but statistics.** Without question, statistical tools and measurements are essential in this methodology, but it's not the end! There are many other external and internal aspects to it, such as putting the customers first, but also stimulating the necessary cultural changes within the organization or department.

Conclusion

In this book, we took a look at the most important elements of the famous Lean Six Sigma methodology. In this book, we took a look at two parts. The first being The Essentials and the second being The Process. These phases were filled with various practical chapters to get yourself familiarized with the essential concepts to run Lean Six Sigma projects. The first part, The Essentials, was covered in Chapters 1-4. You learned more about how Lean came about at Toyota, and you learned it's principles and concepts. Some of its powerful principles are: Specify Value, Identify the Value Stream, Create Flow, Pull, and Pursue Perfection. These principles help in eliminating all forms of wastes, namely: Transport, Inventory, Movement, Waiting, Overproduction, Overprocessing, and Defects.

Also, we delved more in-depth in the various roles present in the Lean Six Sigma approach. The multiple roles are related to different colored belts to indicate responsibility levels and tasks. The roles I addressed were: the leadership team, sponsor, implementation leader or champion, the coach, and team member, and process owner. The belts discussed were: The Yellow Belt, Orange Belt, Green Belt, Black Belt, and Master Black Belt Afterward, we took a look at the methodology, countless benefits, and possible training and certifications you can get.

Implementing Lean Six Sigma comes with numerous benefits. Lean Six Sigma includes the integration of the speed of Lean and the quality of Six Sigma. Six Sigma improves quality (error reduction) by better understanding the business process and ensures a balanced infrastructure. Lean stands for speed by eliminating steps in processes. Both are needed to reduce the costs of complexity in business processes. An essential aspect of Lean Six Sigma is process improvement. According to the underlying philosophy, a group of professionals is needed to effectively and efficiently solve problems.

Above all, a team of professionals, trained in aspects of Lean Six Sigma, works on a problem that is relevant to the organization and its customers. Lean Six Sigma strives for higher quality in less time. To achieve that objective, the organization and process flows of a company must be clear, and everything that is unacceptable to the customer must be eliminated. The people working in different processes can deliver more in a team and share expertise and ideas to solve a problem. All decisions are based on data and facts.

Lean Six Sigma can help a manager achieve goals such as cost savings, more efficient use of the budget, and better customer satisfaction in a structured and substantiated manner. Lean principles are applied to get speed in the processes. As a result of this, it's necessary to recognize waste. Statistics also adds something to process improvement. Is there a visible trend in a process, or is it just coincidence? The following applies to Lean Six Sigma: we intervene only when it's necessary because intervention often costs extra time, energy, and money and can harm the morale of the team. It must also be possible to quantify relationships, for example, between lead times and the number of ongoing projects. To improve, the process must first be understood. By using the DMAIC cycle, we can do the same.

In the second part of this book, The Process, we took a more detailed look at the most critical process surrounding Lean Six Sigma, namely the DMAIC cycle. We took a look at every phase in the cycle and essential tools and techniques. For the Define Phase, we took a look at the Voice of the Customer and Process Mapping with the SIPOC technique. For the Measure Phase, we learned more about data and defects. Furthermore, we learned about finding root causes by using techniques such as the Five Whys method in the Analyze Phase. When the Analyze Phase was cleared, the Improve Phase helped us generate solutions to tackle the root causes we found. Finally, with techniques such as Poka Yoke, it's possible to keep control of the process after the issue was fixed. Furthermore, we learned that applying Lean Six Sigma doesn't happen in a vacuum. Instead, it happens by being much involved with your customers and team members. These are precisely the elements that are present in agile methodologies, such as Scrum. Thus, it's possible to incorporate various aspects of the Scrum approach with Lean Six Sigma.

There are also mistakes you can better avoid for any worthwhile success. When you avoid analysis paralysis, don't get overwhelmed, don't just draw conclusions, don't think that you "know it already," and stop embracing false beliefs regarding Lean Six Sigma, you're well on your way to creating, keeping, and delivering more valuable projects as a project manager!

Check out another book by Robert McCarthy

Printed in Great Britain
by Amazon